Science and Technology Development in the Gulf States: Economic Diversification Through Regional Collaboration

The Gulf Research Center Book Series at Gerlach Press

The GCC in the Global Economy
Ed. by Richard Youngs
ISBN 9783940924018, 2012

Resources Blessed:
Diversification and the Gulf Development Model
Ed. by Giacomo Luciani
ISBN 9783940924025, 2012

GCC Financial Markets: The World's New Money Centers
Ed. by Eckart Woertz
ISBN 9783940924032, 2012

National Employment, Migration and Education in the GCC
Ed. by Steffen Hertog
ISBN 9783940924049, 2012

Asia-Gulf Economic Relations in the 21st Century:
The Local to Global Transformation
Ed. by Tim Niblock with Monica Malik
ISBN 9783940924100, 2013

A New Gulf Security Architecture:
Prospects and Challenges for an Asian Role
Ed. by Ranjit Gupta, Abubaker Bagader,
Talmiz Ahmad and N. Janardhan
ISBN 9783940924360, 2014

Gulf Charities and Islamic Philanthropy
in the 'Age of Terror' and Beyond
Ed. by Robert Lacey and Jonathan Benthall
ISBN 9783940924322, 2014

State-Society Relations in the Arab Gulf States
Ed. by Mazhar Al-Zoby and Birol Baskan
ISBN 9783940924384, 2014

Political Economy of Energy Reform:
The Clean Energy-Fossil Fuel Balance in the Gulf
Ed. by Giacomo Luciani and Rabia Ferroukhi
ISBN 9783940924407, 2014

The Silent Revolution: The Arab Spring and the Gulf States
Ed. by May Seikaly and Khawla Mattar
ISBN 9783940924346, 2014

Security Dynamics of East Asia in the Gulf Region
Ed. by Tim Niblock with Yang Guang
ISBN 9783940924483, 2014

Islamic Finance: Political Economy, Performance and Risk
Ed. by Mehmet Asutay and Abdullah Turkistani
ISBN 9783940924124, 3 vols set, 2015

Gulf Research Centre Cambridge
Knowledge for All

Employment and Career Motivation in the Arab Gulf States: The Rentier Mentality Revisited
Ed. by Annika Kropf and Mohamed Ramady
ISBN 9783940924605, 2015

The Changing Energy Landscape in the Gulf:
Strategic Implications
Ed. by Gawdat Bahgat
ISBN 9783940924643, 2015

Sustainable Development Challenges
in the Arab States of the Gulf
Ed. by David Bryde, Yusra Mouzughi and Turki Al Rasheed
ISBN 9783940924629, 2015

The United States and the Gulf:
Shifting Pressures, Strategies and Alignments
Ed. by Steven W. Hook and Tim Niblock
ISBN 9783940924667, 2015

Africa and the Gulf Region:
Blurred Boundaries and Shifting Ties
Ed. by Rogaia Mustafa Abusharaf and Dale F. Eickelman
ISBN 9783940924704, 2015

Rebuilding Yemen: Political, Economic and Social Challenges
Ed. by Noel Brehony and Saud Al-Sarhan
ISBN 9783940924681, 2015

Science and Technology Development in the Gulf States:
Economic Diversification through Regional Collaboration
Ed. by Afreen Siddiqi and Laura Diaz Anadon
ISBN 9783959940023, 2016

Higher Education Investment in the Arab States of the Gulf:
Strategies for Excellence and Diversity
Ed. by Dale Eickelman and Rogaia Mustafa Abusharaf
ISBN 9783959940122, 2016

Iran's Relations with the Arab States of the Gulf:
Common Interests over Historic Rivalry
Ed. by Maaike Warnaar, Luciano Zaccara and Paul Aarts
ISBN 9783959940047, 2016

Egypt and the Gulf: A Renewed Regional Policy Alliance
Ed. by Robert Mason
ISBN 9783959940061, 2016

Intellectual Property Rights:
Development and Enforcement in the Arab States of the Gulf
Ed. by David Price and Alhanoof AlDebasi
ISBN 9783959940108, 2016

The Arab States of the Gulf and BRICS:
New Strategic Partnerships in Geopolitcs and Economics
Ed. by Tim Niblock, Alejandra Galindo and Degang Sun
ISBN 9783959940085, 2016

The Arms Trade, Military Services and the Security Market in the Gulf States: Trends and Implications
Ed. by David B. Des Roches and Dania Thafer
ISBN 9783959940160, 2016

Science and Technology Development in the Gulf States: Economic Diversification Through Regional Collaboration

Edited by Afreen Siddiqi
and Laura Diaz Anadon

Gerlach Press

First published 2017
by Gerlach Press
Berlin, Germany
www.gerlach-press.de

Cover Design: www.brandnewdesign.de, Hamburg
Printed and bound in Germany by Hubert & Co, Göttingen
www.hubertundco.de

British Library Cataloguing in Publication Data.
A catalogue record for this book is available from the British Library.

Bibliographic data available from Deutsche Nationalbibliothek
http://d-nb.info/1103242059

ISBN: 978-3-95994-002-3 (hardcover)
ISBN: 978-3-95994-003-0 (ebook)

Contents

Acknowledgements

We thank the organizers and sponsors of the fourth annual Gulf Research Meeting, held at the University of Cambridge in August 2014, during which we conducted our workshop, "Science and Technology Education, Research, and Innovation in GCC Countries" that provided the basis for this book. We are grateful to Dr. David Hajjar of Weill Cornell Medical College for his input in planning this workshop, and special thanks goes to Sanya Kapasi and Elsa Courdier of the Gulf Research Center for providing exceptional logistical and administrative support. We thank the workshop participants for their lively engagement, discussion and thoughtful feedback on many of the papers included in this volume.

The participants included Dr. Saad Bakry (King Saud University, Saudi Arabia), Dr. Samir Al-Baiyat and Dr. Iyad Al-Zaharnah (King Fahd University of Petroleum and Minerals, Saudi Arabia), Dr. Halim Redhwi and Dr. Ian Proctor (Dhahran Techno Valley, Saudi Arabia), Dr. Salah Al-Mazidi (Kuwait Institute for Scientific Research), Dr. Fahed Al-Sumait and Marta Tryzna (Gulf University of Science and Technology, Kuwait), Aruba Khalid (Abu Dhabi, United Arab Emirates), Dr. Khaled Machaca, Dr. Edward Hillhouse, Sadaf Lynes, and Sheila Qureshi (Weill Cornell Medical College - Doha, Qatar), Dwaa Osman (Georgetown University in Qatar), Kristin Wu (National Environment Agency, Singapore), Dr. Rakhi Rashmi (Kings College, London), and Dr. Ashley Kranov (Washington State University, USA).

The papers from the workshop that focused on the role of international partnerships to build science and technology capacity in the GCC are included in this volume, and we have included additional chapters from other researchers working on important related areas. We thank the Middle East Initiative and the Kuwait Program at Harvard Kennedy School for funding our multi-year research project on *Science, Technology, and Innovation in the GCC.* The funding from this grant initiated our research work in this area, and has allowed us to explore the multi-faceted issues in the nascent science and technology enterprise in the region.

Afreen Siddiqi
Cambridge, MA, USA

Laura Diaz Anadon
London, UK

May 2016

1

Introduction
Developing Science and Technology in the New Century: International Collaborations in the GCC

Afreen Siddiqi and Laura Diaz Anadon

The notion of knowledge-based economies has gained currency in the state capitals of the Gulf Cooperation Council (GCC). Over the past decade, policy documents and vision plans have laid out roadmaps for a future in which GCC citizens are highly educated and embedded in globally competitive domestic enterprises, and where oil and gas exports are not the sole but a part of the mix driving the national economic engines (ADEV 2008; QNDS 2011; MEP 2016). The visions conjure a future that would be radically different from the region's past.

These aspirations build on the gains that have been made in the GCC on the education front. Primary and secondary education access has risen (WB 2007), and public investments enabled by almost a decade of sustained budget surpluses (QNBK 2015) have sought to catalyze higher education.

While progress in educational attainment is an important milestone, many challenges remain. For instance, the quality of public primary and secondary education, development of communication and analytical skills, and performance in mathematics and science is severely lagging as compared to international standards (TIMSS 2011). This trend manifests through in the quality of higher education in science and engineering in the region.

National governments are attempting to stimulate new sectors of economic activity to provide opportunities for the burgeoning working-age population. The need to diversify GCC economies from the exploitation of non-renewable fossil resources is partly driven by the demographic balance, which has tilted heavily towards the young. An important element linked to goals of economic diversification is development of domestic science, technology, and innovation (STI) capacity (Smith and Abouammoh 2013). As a result, a host of new universities, education and training programs, research centers, and science and technology parks have emerged in the GCC countries.

The attempts to advance STI capabilities mark a new phase of development for the region. After the discovery of oil in the late 1930s and the ensuing economic transformation of the region, efforts for acquiring technical capability and know-how did not commence immediately after. The first universities were established in the late 1950s, but scientific research and the promotion of local innovation capability were not high priorities at that time. The modus operandi between the 1950s and the late 1990s involved importing technology, awarding turn-key projects, and attracting expatriate labor and skilled professionals to run the engine of industrial activities (Zahlan 2012)[1]. Although there was some gradual progression over the decades, the level of local scientific research and innovation remained limited until the late 1990s. Universities served primarily to teach and train – most offered only undergraduate level degree programs in technical and science areas – and research was a lower priority.

Driven by an array of government actions and public-private partnerships, the missions and visions of science and technology (S&T) institutions in the Gulf are now undergoing significant change. The founders of new universities explicitly call for embracing the Humboldtian model – wherein universities are not merely expected to be repositories of knowledge but also produce new knowledge through research pursuits (Schimank and Winnes, 2000).

Saudi Arabia allocated, 56 billion US dollars, a quarter of the total national budget (excluding defense spending) for 2014 for education (JADWA 2014). Of this budget, 10 billion dollars were for higher education. The Research and Development (R&D) expenditure is now almost 1% of the national GDP – well below what other industrial countries spend – but is representative of efforts in expanding research activities. Ten years ago, in 2005, the R&D investment stood at 0.4% (TE 2016). Qatar has made significant investments, including establishing an Education City in Doha, and a multi-billion dollar Qatar National Research Fund (that serves to provide funding for scientific research projects in the country). In the United Arab Emirates (UAE), Mubadala, the state-owned investment company, has made investments in aluminum, aerospace, semi-conductors and clean energy (including solar and wind power) projects (MBD 2016) as part of efforts for industrial diversification and for acquiring technical capabilities in these sectors for future development. It has also created Masdar city as a test bed for new technologies for sustainable cities and a new graduate research Institution, the Masdar Institute. In Oman, the government allocated 13% of the 2013 budget to education, spent 0.5% of its GDP on research, and with its depleting oil wells is at the forefront in utilizing sophisticated enhanced oil recovery technologies in the region (TWAS 2015).

Driving and supporting most of these initiatives are partnerships, collaborations, and contracts, with international and domestic actors including universities, corporations and international organizations. In this book, we take stock of the state of science and technology enterprise in the region through the lens of international research collaboration, models of institutional partnerships, and broader socio-economic issues that both advance and hinder S&T in the GCC. We explore the emergent trends in STI in the Gulf in two parts of this edited volume. The first part analyzes patterns of international STI research collaborations and revealed patterns of affinity across countries, the challenge of attracting and retaining top talent in S&T within the context of international competition, and the newly emerging and comparative role of women in engineering in the region. The second part of the volume presents in detail selected cases of institutional partnership models that have emerged. The authors of these cases are part of or have closely observed the establishment of these organizations. This section is thus a window into the perceptions and experiences of those who have closely witnessed and worked in shaping the international collaborative endeavors.

As a whole, this volume documents and provides insights about the practices and models that have been employed for building science and technology capacity in this century through international collaborations in the GCC. Some features that emerge suggest that the region's cultural, socio-economic, and demographic characteristics may be giving rise to a type of development approach that is characterized by a high level of 'outsourcing', as demonstrated by the relatively high reliance on international workers in S&T, but also with a greater integration of women in S&T at the higher education level compared to industrialized countries. The three international collaborations presented here cover different scopes, ranging from a research, teaching and commercialization effort with an energy and sustainability focus in Abu Dhabi, a biomedical research organization in Qatar, and a medical training institution in Bahrain. By systematically discussing the challenges and tradeoffs of each case, as well as the key importance of the national context, actors, and visions, we hope that this volume will serve as a useful reference as well as a contribution towards understanding how international collaborations are shaping science and technology development in the GCC.

In the rest of this introductory chapter we: (1) provide more detail on the developments in the STI enterprise in the Gulf; (2) summarize key insights from Part I on international partnerships and comparisons; (3) summarize insights from the in-depth cases of international STI partnerships; and (4) conclude with an outlook on the future of STI in the Gulf.

Shifting Landscape of Science and Technology in the Gulf

The initiatives and trends in the STI enterprise in the Gulf since roughly the turn of this century demarcate a shift in science and technology in the region. The establishment of the Qatar Foundation in Doha, in 1996, marked the initiation of a series of education reform and revitalization efforts in Qatar. Weill- Cornell Medical College of New York and Texas A&M University in College Station, Texas opened branch campuses in Doha's new Education City – built and financed by the Qatar Foundation. This trend, of 'internationalization of education' (Witte 2010), proliferated almost simultaneously through the GCC states and continued for a decade. The Royal College of Surgeons Ireland opened its campus in Bahrain in 2004, while NYU-Abu Dhabi's inaugural class started in 2010. In addition to branch campuses of foreign universities, new institutions were established focusing on S&T including Masdar Institute of Science and Technology in Masdar City in Abu Dhabi in 2007, and King Abdullah University of Science and Technology in Saudi Arabia in 2009. The initiatives also included scholarship programs for sending large numbers of GCC nationals abroad for higher education and training, and major efforts for hiring faculty and researchers to fill the open positions in the expanding local S&T sector (Schiermeier 2013).

The billions of dollars of investments were in the backdrop of sustained high oil prices that contributed to budget surpluses for national governments[2]. But as oil prices resume their historical patterns of rise and fall, the volatility of this commodity lays bare questions of building STI institutions for lasting change.

Comparative Perspectives on Key Challenges

The set of chapters in Part I (Chapters 2-4) provide a cross-country comparison of Gulf countries and other selected countries (of North America, Europe and East Asia) to explore the evolution of scientific publications and international co-authorship patterns, workforce development in STI through attraction and retention of international talent, and women's participation in higher education in engineering in the GCC. The cross-country perspectives paint a picture of common challenges that arise from shared cultures and structures of the economy, but also of differences among the GCC states.

In the first chapter, we set the stage by highlighting how productivity (scientific journal publications per capita) trends have evolved, and show that although some gains have been made, the output of Gulf countries remains at less than 1% of global share. We also conduct an analysis of co-authorships in scientific publications using a collaboration affinity metric that provides a measure of extent of collaboration while controlling for size of publication output of countries. This analysis shows that the GCC states have developed stronger relationships among themselves in terms of joint-publications over time. In addition, scientific publication affinity between GCC and Egypt is strong, and furthermore, new affinity between researchers in GCC states and Maghreb countries has emerged in the 2000-2010 decade. The results as a whole reveal patterns of integration – as measured through collaboration between individual researchers – as well as of extension and outward connections with other regions. There are differences among the GCC states: Saudi Arabia shows greater links with regional countries of the Middle East and North Africa (MENA), whereas Qatar and UAE lead with affinity for countries not only in MENA but also outside the region including North America, Europe and Asia. Overall, this chapter shows the evolution of scientific research partnerships in a region undergoing significant STI institutional change in real-time, highlighting the important role of cultural affinity and geographic proximity in the past decades.

Aruba Khaled's chapter on "Building a Science, Technology and Innovation Workforce" provides an overview of the challenges related to attracting large fractions of migrants in the Gulf. Countries in the Gulf area are relatively unique in terms of the migrant population as a fraction of total population. For example, migrants accounted for 84 per cent of the UAE's population in 2013, making it the top ranked country in terms of international migrant stock. Other Gulf countries are also highly ranked, with Qatar following at third, Kuwait 7th, Bahrain 13th, Saudi 33rd and Oman 34th in hosting international migrants. After discussing some of the factors driving the high dependency on foreign workers in GCC countries (e.g. small native population, low labor force participation, initially low literacy and education levels, and high non-labor income for nationals), this chapter compares various Gulf countries in terms of their temporary (or guest) working contracts and labor productivity. It argues that temporary contracts are resulting in a churn of talent that limit its potential as a top destination for highly educated and qualified people in the STI sectors. Furthermore, it argues that the region still does not have sufficient local capacity, highlighting even further the need for government action in this area. Providing a desirable environment for talented workers is, of course, not only a challenge in the Gulf. The chapter uses experiences of industrialized countries, including Germany's Green Card Initiative, Australia's General Skilled Migration Program, and Canada's Federal Skilled Worker program to describe routes that have been pursued at state-level to facilitate international talent recruitment and retention. It concludes by highlighting the importance of labor and immigration policies in the development of the science, technology and innovation enterprise in these countries.

DeBoer and Kranov's focus on the achievement of females in science and engineering programs in the Gulf elaborates on a key component of the workforce development challenge described in Khaled's chapter. Many will find surprising that across most of the GCC nation-states, women attain degrees at the tertiary level at near parity to male students, and they participate in engineering undergraduate studies at significant rates. However, female participation in the general workforce in the GCC still stands at 27%. One question that deserves further attention is why female participation in higher education in engineering is higher in Muslim countries than in other western countries, including the United States. After reviewing significant differences across GCC countries in terms of participation and attainment, attitudes toward science, technology, engineering and mathematics (STEM) education, the authors perform a comparative analysis of one GCC country (Saudi Arabia) and other countries: Jordan (Middle East), Tunisia (North Africa), Malaysia (another majority-Muslim country in Southeast Asia) and the United States. Many emergent (and so far unresolved) questions include what the United States, for instance, can learn from experiences in the Gulf, since its efforts to revamp female participation in engineering have not yielded significant results. DeBoer and Kranov point to the potential roles of online education to improve attainment and of using ICT technologies to increase female workforce participation, while warning that they could in practice lead to further perpetuation of separation and exclusion.

A Variety of Approaches and Local Contexts to Building Science and Technology Institutions

Part II of this edited volume consists of three case studies that were developed by researchers intimately involved with the creation of three types of entities at different levels: a biomedical research effort at WCMQ, a medical school in Bahrain, and a university incorporating both research and graduate education in Abu Dhabi. Common to these three efforts, which differed in focus, were partnerships between the governments of those countries and internationally recognized science and technology institutions in high-income countries. They span different types of organizations and partnerships, and highlight the different impacts.

Pfotenhauer analyzes the creation of the Masdar Institute of Science and Technology as a co-production of Emirati science and society. The Masdar Institute was founded in 2006 and conceived as "a world-class research and education environment that attracts and nurtures future leaders and critical thinkers in science and technology with particular focus on advanced energy and sustainability" (MIST 2011). The co-production can be exemplified by a continued reliance on "caravans of knowledge" (*i.e.*, expats) that characterize the current society, but also by a vision of what the Emirates could become embracing a knowledge-based more sustainable society. The Masdar Institute currently hosts about 500 students and 90 faculty and has 9 masters programs and a doctoral program. The development of the Masdar Institute was done in collaboration with MIT under a contract aiming to achieve "institutional transfer", and is part of a broader innovation ecosystem that includes Masdar City (to test technologies in a real city environment), Masdar Clean Energy (a corporate development medium to focus on sustainable energy), and Masdar Capital (to fund high risk commercialization projects). The author argues that Masdar is both a product and a shaper of Emirati society, with a focus on Arab architecture and on regional problems. Pfotenhauer argues that the partnership with MIT was essential to the setting up of Masdar. The Institute drew upon MIT's scientific expertise, operational resources, global visibility and credibility, and physical location and the partnership served to stabilize the fragile foundations of knowledge production. Of 189 new students enrolled in 2014, 47% were Emiratis, and many more stem from other Arab-speaking countries. In looking ahead to the future, the author notes that some of the key challenges for Masdar include the need to increase the attendance of local students, and of increasing well-paid domestic jobs in STI sector.

Machaca and Sheikh discuss the design and structural features of the Biological Research Program (BRP) that started in 2009 within the Weil Cornell Medical College-Qatar (WCMC-Q) campus, which was set up in 2001 under a fee with WCMC, to award medical degrees with the same standards as those set up by WCMC. With the goal of creating a globally competitive medical research center in Qatar, the research program has grown to 37 active research laboratories with research interests spanning the basic molecular, translational and clinical research arenas. The growth to 200 staff has also led to over 270 papers and some extramural funding to support their research.

The chapter discusses the infrastructural, leadership, policy and logistical issues that had to be addressed to create the BRP. The program had to build from scratch systems to provide research support administration, expertise in grants management and commercialization, ethical compliance, clinical research support and lab management, in addition to creating core lab facilities. The authors discuss the need to increase the participation of Qatari nationals (which currently make up about 27% of WCMC-Q medical students) and to complement the biomedical research with a graduate level program in the biosciences.

Whitford, Mirza and Lawrence discuss the creation of a different type of medical school campus in Bahrain, the Royal College of Surgeons in Ireland (RCSI) Bahrain. RCSI Bahrain had the goal of addressing shortages in medical care specialists in the country. With lower levels of wealth than some of its Gulf neighbors, Bahrain not only had a different goal, but it also took a different approach to attracting international branch campuses by focusing on building local capacity as a priority over importing expertise and being at the forefront of science. RCSI Bahrain is fully-owned and operated by RCSI Dublin and did not receive funding from Bahrain beyond the land used to build the campus. The authors discuss the importance of the personal relationships between RCSI and Bahrain—which date back to the 1970s, when Bahraini's went to RCSI for surgical training—in the creation of the partnership. There have been 854 graduates since 2010, however, it is still too soon to measure long-term impact after graduation. The authors highlight that there has been little expatriate turnover. Furthermore, RCSI Bahrain has retained a relatively high percentage of the RCSI Bahrain's jobs for the national population and the enrolment of country nationals is at 40% in the college. In reflecting on lessons learnt from this experience, the authors note the need of providing students with training on critical thinking and working on a cultural fit.

Looking Forward to the Future of Science and Technology in the GCC

The GCC countries are considered as newcomers in their efforts to establish a position of strength in technical knowledge and innovation capabilities (Andersson and Djeflat 2013). This offers a unique opportunity to study the ways in which the region is attempting to acquire and establish scientific and technical expertise. The GCC efforts in the STI arena are not only of interest because of the unique situation of the region in terms of wealth and culture, but also because their big push in STI is coinciding with an era of accelerating discoveries and innovations, posing unique challenges and opportunities for nations aiming to achieve competitiveness in research and technology innovation.

Gulf countries have laid down some foundations for developing indigenous capabilities in science and technology. However, a rigorous assessment of the impact and effectiveness of the new institutions, partnerships, centers of research, and scholarship programs will only become possible after several years given the long-term nature of education and capacity development efforts. Nonetheless, the experimentation of the last few decades offer opportunities for learning within the GCC and internationally.

On some metrics there are clear signs of advancement. In 1970, there were seven females enrolled in university courses in Saudi Arabia. By 2011, women made up more than 60 % of all university enrolments, and 25 % of enrolments in master's and doctoral degree programs in the country (Smith and Abouammoh 2013). However, the participation of women in professional engineering, design, and technical management positions remains limited, and it is unclear what their attainment will be in the short term. What is almost certain is that women's higher education is an important forerunner of societal change – one that can have long lasting effect through shaping the sensibilities and attitudes of the future generations to come.

An important factor that may stall or even erase recent STI advances in the GCC is that of dynamism and growth of global science and innovation. Science and technology move at an ever faster pace, so the challenge for nations seeking to join and drive the frontiers of research and innovation is to meet a moving target – not a small feat for countries starting with a low science and technology base. As Andersson and Djeflat aptly note for the MENA region that, "Technology offers opportunities, but also presents risks. Without the required tools and competencies, new jobs will not materialise. Rather, they will shrink farther in number. This is the situation that the Middle East must now deal with." (Andersson and Djeflat 2013). The GCC countries have made considerable efforts to elevate national institutions in international standings and have been attempting to climb up along the international yard-sticks of performance measures. But for continued success, what is needed are new models of education and research that emerge from within -that fit the needs of regional development and particular socio-cultural context—and that enjoy sustained support. Otherwise, the region may always be playing catch up.

The new universities and institutions herald progress, but this interim progress is very dependent on the course of long-term national policies along with global dynamics of the S&T enterprise. The issues of how long the national investments in science and technology will last, given volatile oil prices, and how efficiently the resources can be used to build real capacity linger on given the patterns of the past. If history is to serve as any guide, development of scientific research and technical capability in the post-colonial Arab countries is replete with patterns of initial attempts followed by premature abandonment – often due to wars and strife or due to general ambivalence and disinterest in the sector as a national priority (Zahlan 2012). The continuation of policies and sustained support and investment from the government in concert with public support and interest will be a key defining factor in how S&T grows and gets embedded within the wider economy and society.

References

ADEV. 2008. "The Abu Dhabi Economic Vision 2030", The Government of Abu Dhabi, UAE. Accessed Jan 9, 2016. (https://www.ecouncil.ae/PublicationsEn/economic-vision-2030-full-versionEn.pdf)

Andersson, T and Djeflat, A, Eds. 2013. *The Real Issues of the Middle East and the Arab Spring: Innovation, Technology, and Knowledge Management*, New York: Springer. DOI: 10.1007/978-1-4614-5248-5_18.

Freeman, Christopher. 1987. *Technology Policy and Economic Performance: Lessons from Japan.* London: Pinter.

JADWA. 2014. http://www.jadwa.com/en/download/2014-budget/saudi-arabias-2014-budget.

MBD. 2016. Accessed: January 9. https://www.mubadala.com

MEP. 2016. "The long term strategy of the Saudi Economy". Accessed: Jan 9. (http://www.mep.gov.sa/themes/GoldenCarpet/#1452367247452)

MIST. 2011. Learning to change the world: The Masdar Institute of Science and Technology, Available at: http://www.masdar.ac.ae/data/p7tp9xg71490.pdf.

QNDS. 2011. "Qatar National Development Strategy 2011~ 2016: Towards Qatar National Vision 2030," Qatar General Secretariat for Development Planning, Doha, 2011.

QNBK. 2015. Economic Insight Reports. Qatar National Bank. Accessed: Jan 9, 2016. http://www.qnb.com/cs/Satellite/QNBQatar/en_QA/InvestorReln/Publications/enEconomicInsightReports

Schimank, Uwe and Markus Winnes. 2000. "Beyond Humboldt? The Relationship Between Teaching and Research in European University Systems". *Science and Public Policy*, 27(6): 397-408.

Schiermeier, Quirin. 2013. "Relocating: Middle Eastern Promise," *Nature*, 500: 111-12.

Schumpeter, Joseph, A. 1934. *The Theory of Economic Development.* Oxford University Press.

Smith, Larry and Abdulrahman Abouammoh, Eds. 2013. *Higher Education in Saudi Arabia: Achievements, Challenges and Opportunities*, Dordrecht: Springer

TE. 2016. Accessed: January 9. http://www.tradingeconomics.com/saudi-arabia/research-and-development-expenditure-percent-of-gdp-wb-data.html

TIMSS. 2011. Trends in International Mathematics and Science Study. Available: http://nces.ed.gov/Timss/

WB. 2007. "The Road Not Traveled: Education Reform in the Middle East and North Africa," *MENA Development Report* (The World Bank, Washington, D C, 2007).

Witte, Spencer. 2010. "Gulf State Branch Campuses: Global Student Recruitment", *International Higher Education*, 58: 5-6.

Zahlan, Antoine B. 2012. *Science, Development, and Sovereignty in the Arab World.* New York: Palgrave Macmillan

Notes

1 This pattern followed the traditional regional strength – trade and exchange. For millennia, the commerce and economy in the Arabian peninsula was based on trade and movement of goods. Production of goods and innovation of new products came from the East (Mesopotamia and beyond), North (Levant) and the West (Egypt, Maghreb, and further beyond in Europe).

2 Price of brent crude oil above $60 a barrel from 2006 to 2015 – except for a brief period in 2009. For several years in this period it hovered above $100 a barrel.

2

Science and Engineering Research Collaborations: Patterns of Integration and Extension in the GCC

Afreen Siddiqi and Laura Diaz Anadon

International technical collaboration has been an essential element in the development of the Gulf Cooperation Council (GCC) states. After the discovery of oil in the 1930s, American technology started the change in the fortunes and future of the region. The Arabian peninsula underwent dramatic changes as the oil economy heralded new kinds of international technology partnerships. Foreign engineers, architects, and expatriate labor transformed the region with imported technology into a constellation of new metropolises, skyscrapers, and modern infrastructure. The GCC states – now increasingly aware of the limits of their fossil-fuel powered economies for the 21st century and rapidly changing demographics where the youth now form the majority – are embarking on efforts of economic diversification. One key focus is on building up a local scientific and technological research base to drive future industrial growth. The governments have set up a number of new international partnerships with the aim of strengthening national capacity in research and innovation and gaining a foothold in the global scientific arena.

In this chapter, we analyze research collaboration in the GCC, through a detailed bibliometric analysis of co-authorship data of scientific publications over three decades from 1981 to 2013. The data reflects collaboration between individual scientists and researchers. Joint publications result from a variety of factors ranging from mutual research interest of collaborating scientists to mandatory outputs resulting from bilateral agreements between institutions (Glanzel 2001). International collaborations span different types of channels (such as dual-degree programs, students and teacher exchange, internships, as well as research projects) and publications data is only a partial measure of the extent and type of collaborative activities. Bibliometric analysis, however, uses verifiable and traceable data and a systematic approach for gaining important insights on scientific research output and co-authorship trends. Here, we conduct a long-time series analysis, of data ranging from 1981 to 2013, to characterize how collaborative publications trends have evolved in the GCC and in the wider Middle East and North Africa (MENA) region, how the present patterns compare with the past, and what the implications maybe for future regional development.

An Opportunity to Study the Emergence and Evolution of Research Collaborations in the GCC

The first technical education programs were established in the region in the 1950s, and new institutions continued to emerge through the 1960s and beyond. It was, however, at the transition of this century, when visible and markedly faster pace of government efforts became apparent. International collaborations - with premiere institutions - were a key element through which the policy makers sought to quickly elevate the quality, visibility and stature of some of their flagship efforts[1]. These top-down interventions along with the traditional on-going, bottom-up, researcher-level collaborations have collectively shaped the international engagement of the GCC.

The system-level patterns that can be partly studied with co-authorship data, however, have not been investigated in detail. Several studies and reports have analyzed the new programs, publication volume and emerging citations trends (Smith and Abouammoh 2013; Hajjar 2014; Adams 2012). However, there is limited work on long-term comparative assessment, and on how the collective scientific research sector in the region is emerging and embedding in global scientific enterprise. Here, we characterize international scientific research collaboration in the GCC countries— Saudi Arabia, Qatar, Kuwait, United Arab Emirates, Oman, and Bahrain— using co-authorship data from publications listed in the Science Citation Index - Expanded™ that includes over 8500 leading scientific and technical journals.

Co-authorship patterns and bibliometric studies have long been studied in the European and US scientific research contexts (Glanzel 2001), and continue to be a focus of research that helps understand the structure and evolution of global science (Ma 2015). However, the regional focus on GCC countries has largely not been made. Given the emphasis that has been placed in foreign partnerships, as well as the scale at which such institutions have been set up, one may expect unique aspects in the evolution, nature and structure of scientific collaborations in the GCC region.

In this chapter, we characterize the structure and evolution of international research collaborations by identifying the top partnering countries in co-authorship of journal articles for each GCC state during the decades of 1981-90, 1991-2000, and 2001-2010. We also use a measure of co-authorship *preference*, or 'affinity', that measures the intensity of co-authorship partnerships while controlling for the size of scientific activities (publications) in the collaborating partners. Using this metric, we compare how the GCC states collaborated in scientific publications among each other, and with other countries of the Middle East and North Africa (MENA) region, Asia, Europe and North America.

In the following sections we first provide additional background on the partnerships and collaborative programs in engineering and science education and research in the GCC states, and then review the literature characterizing scientific collaborations in different regions. Next, we present quantitative results of our bibliometrics analysis of publications from the GCC, present results of the top partner countries with which GCC states published scientific and technical articles, and the partnership preferences across various countries and regions. In the last section, we discuss the results and implications of our findings.

Science and Technology Development in the GCC: Initiatives through International Collaborations

International technical cooperation in the GCC dates back to the development of the petroleum and gas sector. The discovery of oil in Bahrain in 1932 by Western petroleum engineers triggered an oil rush, and oil fields were found in rapid succession in Saudi Arabia and Kuwait (in 1938), and in Qatar in 1939. The UAE and Oman followed later in the early 1960s with local oil production (Commins 2012). The ensuing infrastructure development, funded through petroleum exports, was almost entirely based on importation of technology and turnkey projects, with little development of local technical capacity (Zahlan 2012; Andersson and Djeflat, 2013). While local engineering universities and research institutions were steadily established, the national technology development capabilities remained severely limited. There are now evident signs of a shift. Policy makers have taken actions to strengthen thee local industrial base - the extent of such actions, however, vary across the region. Here we briefly list some prominent programs that have been formed with international collaboration in each of the GCC states since the mid 1990s.

Saudi Arabia

One of the most prominent projects in technical higher education in Saudi Arabia was the establishment of King Abdullah University of Science and Technology (KAUST) in 2007 with an endowment of US$10 billion. The Saudi government negotiated US$500 million worth of partnerships with major international universities, including Stanford University, University of California at Berkeley, University of Texas and Cambridge University (Al-Ohali and Burdon 2013). Another flagship initiatve was the King Abdullah Scholarship Program (KASP) established in 2005, and considered to be the largest fully endowed scholarship program ever supported by a nation state (Bukhari and Denman 2013). The program does not have disciplinary restrictions, and funds students pursing degrees in management, social sciences, and science and engineering. In 2014, the annual budget allocated 22 billion Saudi Riyals (~US$ 5.4 billion) to support over 185,000 Saudi students studying abroad (JADWA 2014).

Several other programs and universities have been established in the country since 2005, and major reforms and new projects have been created in existing institutions (including at King Saud University in Riyadh, King Fahd University of Petroleum and Minerals in Dhahran, and King Abdul Aziz University in Jeddah). Furthermore, several technology parks including Dhahran Techno-Valley at KFUPM (in close proximity to the headquarters of Saudi Aramco), and King Abdullah Economic City near KAUST have been created (Al-Sultan 2012). This new phase involves partnerships with multi-national enterprises in the oil and gas, chemicals, and other related sectors (including Dow Chemical Company, Sumitomo, Baker Hughes, Schlumberger and several others) to set up their research and development centers in Saudi Arabia in proximity to technical universities. Through such programs, the Saudi educators and administrators hope to stimulate research activities that can train Saudis through foreign collaboration, rather than sales, business, and marketing offices of technology companies as was the practice in the past. The national science and policy agency – King Abdulaziz City for Science and Technology – has negotiated several agreements with technology companies in energy, aerospace, remote sensing (DoS 2015; KACST 2015) and other strategically identified sectors.

United Arab Emirates

The UAE has drawn up a partnership list with a stellar group of research universities in the US. The array of collaborative programs span technology, healthcare, business education and leadership training (UAEE 2015). Within the science, technology and health care areas some well-known universities include: New York University (which has set up a campus in Abu Dhabi and offers undergraduate level liberal arts and science degrees), Boston University (has opened a Dental school with Dubai Healthcare City), Massachusetts Institute of Technology (has worked with Abu Dhabi Future Energy Company to help establish Masdar Institute of Science and Technology), Rochester Institute of Technology (has established a campus in Dubai), New York Institute of Technology (NYIT), and Colorado School of Mines (has a ten year agreement with Abu Dhabi National Oil Company for development of the Petroleum Institute in Abu Dhabi).

The government also actively sought and succeeded in bringing the world headquarters of the International Renewable Energy Agency (IRENA) in the UAE. It is located in Abu Dhabi in Masdar City–an ambitious project launched in 2008 to establish a modern urban hub for new technologies in energy, transportation, and urban construction (Reiche 2010). The UAE has also made a number of partnership agreements with major international firms. These include an agreement on partnerships in clean energy research, aviation, and commercial finance between General Electric and the state-owned Mubadala Development Company. Boeing and Mubadala Aerospace have a long-term agreement, and Strata Manufacturing (a Mubadala Aerospace company), has a 10-year direct contract for composite aero-structures for Boeing 777 and 787 Dreamliner aircraft. The semiconductor manufacturer Advanced Micro Devices (AMD) and Abu Dhabi's Advanced Technology Investment Company (ATIC), have created a semiconductors manufacturing company headquartered in the US. In the information technology sector, Injazat Data Systems, is a joint venture between Mubadala Development Company and Texas-based Electronic Data Systems, that provides business process services in the Middle East region. The government has also set up a Technology Development Committee (TDC), IP regulation, and free zones as parts of a strategy to kindle economic activity and incentivize foreign technology firms.

Qatar

Qatar led the region in the push for education and a vision for establishing a 'knowledge economy', with the establishment of Qatar Foundation (QF) in 1996. The Qatar Foundation has the mandate for education, science and research, and community development and has been the driving vehicle through which other initiatives were launched in the country. It launched Education City in Doha in 1997 and helped in bringing eight foreign university campuses (six from the US, and one each from UK and France). These include Weill Cornell Medical College in Qatar (established in 2001), Texas A&M University at Qatar (established in 2003), and Carnegie Mellon University in Qatar (established in 2004) that provide medical, engineering, and computer science programs respectively. These branch campuses are largely focused on education and training, however research is part of their agenda and research activities have grown steadily. Texas A&M at Qatar valued its research activities at reaching approximately $200 million (TAMU 2105). The Qatar Foundation set up Qatar National Research Fund (QNRF) in 2006 to serve as a funding agency for research focused on strategic areas of national interest (including energy and environment, computers sciences and information technologies, and health and life sciences). The QF has also set up the Qatar Science and Technology Park that aims to provide a platform for applied research, innovation and entrepreneurship. It provides a 'Free Zone' for entrepreneurs to set up a local company or a branch of a foreign company with no tax obligations, no requirements of Qatari partnership (100% foreign ownership is allowed) and duty free imports (QSTP 2015). In 2009, QF launched the World Innovation Summit for Education (WISE) that has achieved international visibility and annually convenes large number of educators for conferences, and other programs.

Kuwait

Kuwait has had a long history of international scientific engagement. Kuwait Institute for Scientific Research (KISR), the country's primary research institution, was established in 1967 with Japanese collaboration as part of an agreement with Japan's Arabian Oil Company for an oil concession. With an initial focus on applied research in petroleum, desert agriculture, and marine biology, the institute has considerably broadened its disciplinary areas over the years (KISR 2015). Since the 1990s, a number of international experts, institutions and private consultants were hired for strategic planning for economic development (Brinkley 2012), and there has been a steadily growing emphasis on strengthening STI in the country with some notable efforts in the last decade. KISR's 7th strategic plan (for the 2010-2015 period) aimed to over-haul the planning process and reform the organization, and Kuwait University – the largest university in the country – specifically includes scientific research as a component of its primary mission (KU 2015).

The Kuwait Foundation for Advancement of Science (KFAS) - the main research funding agency in the country since 1976 – has embarked on a number of agreements with leading foreign institutions (KFAS 2012). Some key programs include a Kuwait Center for Natural Resources and Environment at the Massachusetts Institute of Technology (established in 2005 for collaborative research and education in energy, water, and environment (CNRE 2015)), the Kuwait Program at Sciences Po (a cooperative agreement made in 2007 and renewed in 2012) with the Fondation Nationale des Sciences Politiques (Sciences Po) in France for joint initiatives in research, teaching, and executive education, Kuwait program at London School of Economics on Development, Governance, and Globalisation in the Gulf States (LSE 2016), and the Kuwait Program at Harvard University for research, teaching and training (HKS 2015).

There are some private institutions such as Gulf University of Science and Technology (GUST), established in 1997, serving a niche role in meeting the growing demands for higher education in the country. GUST has a dual enrollment agreement with University of Missouri- St. Louis in the US, and a new academic-industry partnership has also been formed between GUST, UMass Lowell and Raytheon Co. (UML 2015). Additional private institutions include American University of Kuwait (established in 2003), Australian College of Kuwait (started in 2004), and American University of the Middle East (founded in 2007) that offer degree programs in engineering and computer science.

Oman

Oman started efforts for expanding local technical capacity in the 1980s. The main education institution, Sultan Qaboos University established in 1986, offers higher degrees in science, engineering, medicine and health and is the largest university in Oman. It is also the primary research institution in the country. The largest private engineering college in the country, Caledonian College of Engineering, was established in 1996. It is affiliated with Glasgow Caledonian University in the UK and Vellore Institute of Technology in India and has a number of international research partnerships including some with Australian and Indian institutions (CCE 2015).

In the last decade, a number of new institutions have been created including the first maritime college in the GCC, International Maritime College Oman, which was established in 2005 by the Government of Oman with STC Group in Rotterdam in the Netherlands. It offers degree programs in nautical studies, marine engineering, and shipping, port and transport management (IMCO 2015). In the same year, the government also set up the Research Council as a policy-making body and research-funding agency to promote research, innovation and science (TRC 2015). Two years later, the German University of Technology in Oman was established in 2007 (GUTech 2015), and in 2011, Sultan Qaboos decreed the establishment of a new government university, University of Oman, dedicated for education and research in science and technology. The university campus along with a Science and Technology City in Muscat is planned to provide space for international research institutes and start-up ventures (ODO 2015).

Bahrain

Bahrain is a center of finance and banking in the region, and has a developed tourism sector. The national investments have been less in expanding science and technology and an industrial base as compared to some other GCC states. One major international collaborative project has been the establishment of the Royal College of Surgeons Ireland (RSCI) in Bahrain in 2004. New York Institute of Technology (NYIT) established a NYIT-Manama in 2003, however it closed its campus in 2014 (NYIT 2015) due to changes in regulations for private institutions.

Research on International Co-authorship Patterns

International collaborations are an important contributor to knowledge transfer for countries that aim to acquire new technical capabilities. Such collaborations are an important pathway for bringing new programs at par with global standards, attracting qualified faculty and well-prepared students, and for quickly learning about contemporary scientific problems at the frontier (Katz 1996). As the GCC countries have embarked on creating local institutions for scientific research and have sought international collaborations to aid and elevate these efforts, an examination of international co-authorship patterns in the GCC countries can provide valuable insights on the emergent scientific activities.

Collaboration (and co-authorship) is intrinsic to scientific research, and has long been the subject of study. In 1760, ~ 2% of the scientific papers were co-authored, and by the end of the 20th century this fraction had grown to 80% of publications in the natural sciences (Godin and Ippeseil 1996). The rate of co-authorship has risen in parallel with rising fractions of international co-authorships (where scientists resident in different countries collaborate and jointly author papers). The results of a number of studies show that international collaborations lead to higher visibility and citation rates (Adams 2013), and ever larger teams are working on grand challenges in physics, environmental sciences and life sciences (Wuchty 2007).

Several researchers have applied graph theory to study co-authorship networks (Newman 2001; Barabasi 2001). This method has also lent itself well for studying structure of scientific communities and investigate implications of topology on knowledge flow (Sorenson 2006), rate and extent of innovation, and effects on citations and impact. Most of these studies have sought to examine features of co-authorship networks to elicit structural insights such as size and centrality.

The specific international co-authorship patterns have been found to be aligned along geographic proximity, and cultural, language, and historical (such as colonial) connections (Schubert and Glanzel 2006; Archambault 2011). In case of national technical capacity development, however, there has been less research on determining correlations between patterns of international co-authorship and level of 'catch-up' or advancement in scientific capacity. As a first step towards understanding how the research activities are shaping up in the GCC and how the community is linked to external researchers, a meta-analysis of the patterns of scientific output (measured through annual publications over time) and affinity (measured through co-authorship data) allows for characterizing these key features.

Shifting Patterns in Scientific Publications in the GCC

We analyzed publications records in science and engineering journals sourced from the Core Collection of Science Citation Index-Expanded from Web of Science (WoS 2015). English is the dominant language for scientific publications in the GCC, so we only included English language journal articles, and analyzed records of articles that were published between 1981 and 2013.

Figure 1 shows the number of journal articles published annually in which at least one author had an address from an institution in Bahrain, Oman, Kuwait, Saudi Arabia, Qatar, or the UAE[2]. We find that there was little publication activity in the early 1980s, and that the largest number of publications were associated with researchers with addresses in Saudi Arabia. During the 1990s, there were periods of stagnation and even decline in output from that country. However, since the mid-2000s the output has skyrocketed increasing from 1319 in 2005 to 8800 journal articles in 2013. Kuwait hosted the second largest research activity in the region in the 1980s. Its upward trend was arrested with the Gulf war in 1991 after which it took the country almost a decade to reach its pre-war publications output level and its pace of growth has not increased. The output has steadily increased in the UAE, Oman, and Qatar – and the rate of growth has been the fastest in Qatar since the mid 2000s. Bahrain has not shown any change in rate of output over the period of analysis. Overall, there has been an increase in research publications in science and engineering fields in the GCC, with noticeable shift in pace of growth since the mid-2000s in Saudi Arabia and Qatar.

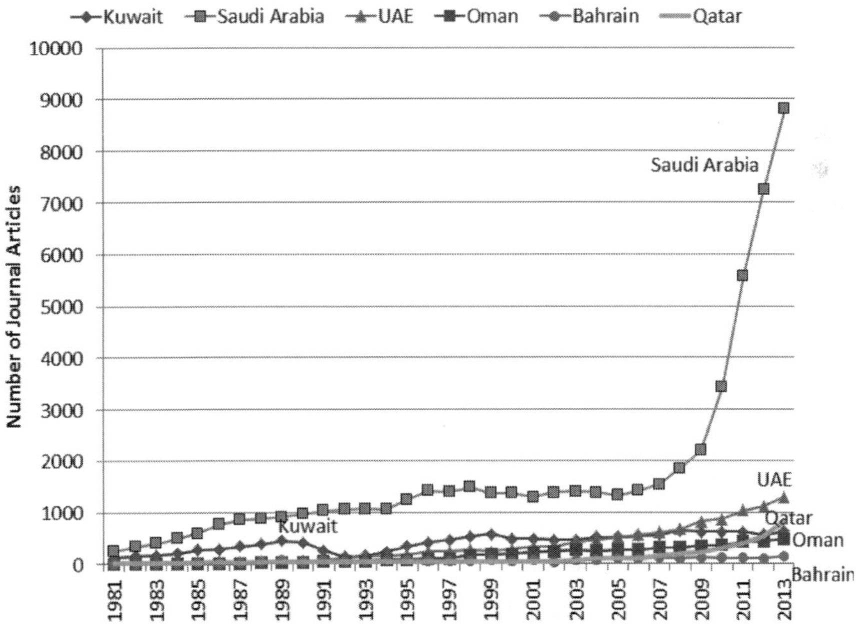

Figure 1: Total numbers of science and engineering publications with at least one author with one address in the GCC between 1981-2013.

The productivity (publications divided by population of each country[3]) also shows changing trends. Figure 2 shows the results using the publications data and the annual population estimates for each country. We find that journal papers published per million people were highest in Kuwait among the GCC countries. The large number of clinical research and medical publications from the country put it ahead of the other states. However, efforts in Qatar and Saudi Arabia over the past 10 years have had impact, with the productivity from those two countries surpassing that of Kuwait in 2011. We also compared the publications in the six GCC states with that of eleven other countries of the Middle East and North Africa (MENA): Morocco, Libya, Algeria, Tunisia, Sudan, Egypt, Jordan, Syria, Lebanon, Yemen, and Iraq. We find that GCC states have had higher publication productivities when compared with the MENA average. The MENA average was computed by including data of all 17 countries – six GCC states and 11 other countries. In particular, the productivity of Qatar, Saudi Arabia and Kuwait ranged from 70% to 210% greater than that of the MENA average.

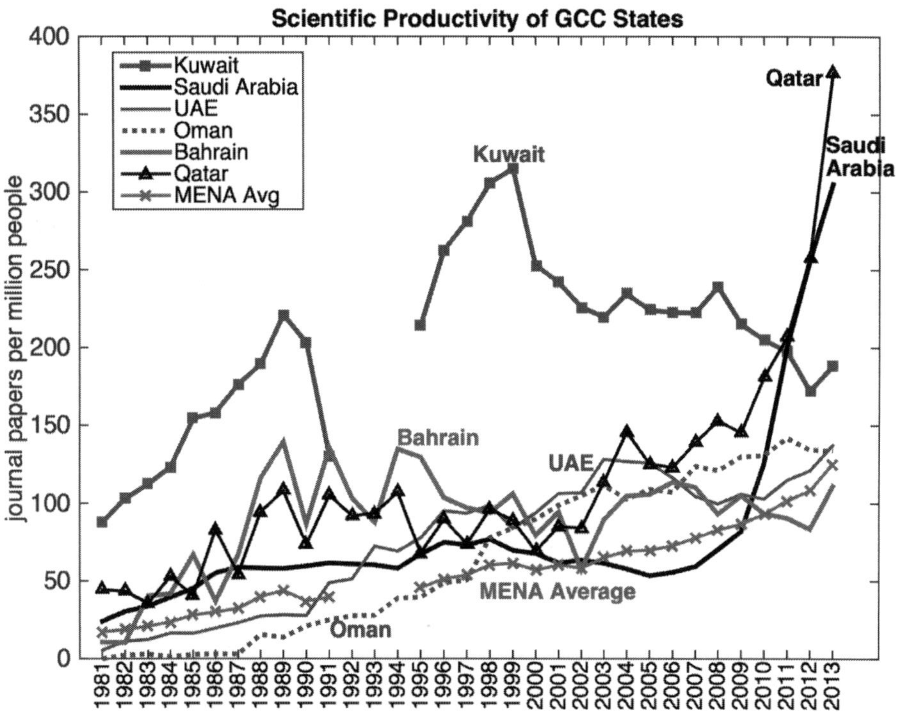

Figure 2: Productivity in the GCC states is higher than the MENA average. There have been notable shifts since 2011 with Qatar and Saudi Arabia surging ahead.

While the GCC have made progress, other countries have too – the scientific research publications have grown globally. When we determine the share of GCC publications within the total global output, we find that the combined share of all six GCC states changed from 0.02% in 1991 to 0.75% in 2013 (see Figure 3). In comparison, however, the other 11 MENA countries had a larger share (1.05%), and so did Iran (1.53%) and Turkey (1.43%)[4]. Figure 3 shows the combined share of all GCC states (of 0.75%) in global publications in 2013.

The gains in publications volume and productivity in the GCC over the past three and a half decades point to a research system that is expanding. However, a number of other aspects, such as quality and novelty, will ultimately determine the effectiveness and competitiveness of the regional scientific sector in the international arena.

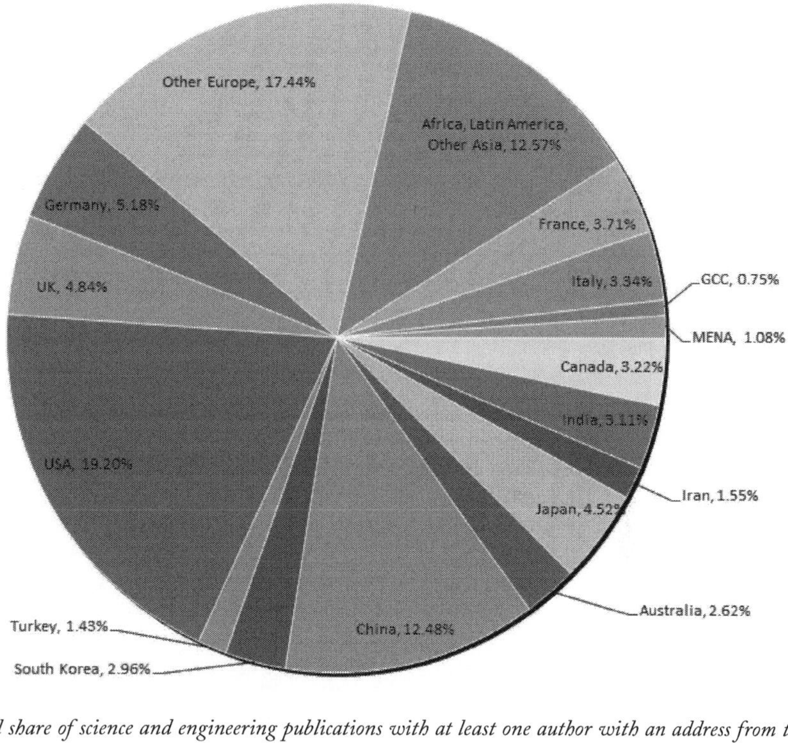

Figure 3: Global share of science and engineering publications with at least one author with an address from the various countries and regions represented in 2013. MENA in this figure includes combined publications from Morocco, Libya, Algeria, Tunisia, Sudan, Egypt, Jordan, Syria, Lebanon, Yemen, and Iraq. GCC consists of Saudi Arabia, UAE, Qatar, Oman, Kuwait, and Bahrain. 'Other Europe' and 'Other Asia' are all countries of those continents except for those shown separately in the pie chart.

International Scientific Collaborations of the GCC

As the GCC states have sought international collaboration for acquiring and accelerating scientific research development in their countries, an examination of the geography of co-authorships in scientific publications will be interesting and useful. The national co-authorship data will reflect realities that emerge from the confluence of a host of factors related to motivations and incentives of individual researchers, as well as the top-down measures that have been implemented for fostering international collaborations in the GCC. Researchers are strategic and opportunistic in their search of partners to gain expertise in particular areas, to build a network of collaborators and partners, and/or to leverage various funding opportunities. Also, researchers (particularly expatriates) in the GCC states often collaborate with former colleagues in their home countries or in the countries of their education and training.

We note that co-authorship analysis only provides partial assessment, however, this approach allows for quantitatively comparing across time and between countries how collaboration patterns have evolved.

Top Publications Partners for GCC Countries

The top five countries with most frequent co-authorships for each GCC state are shown in Figures 4-6 for the 1981-2010 period[5].

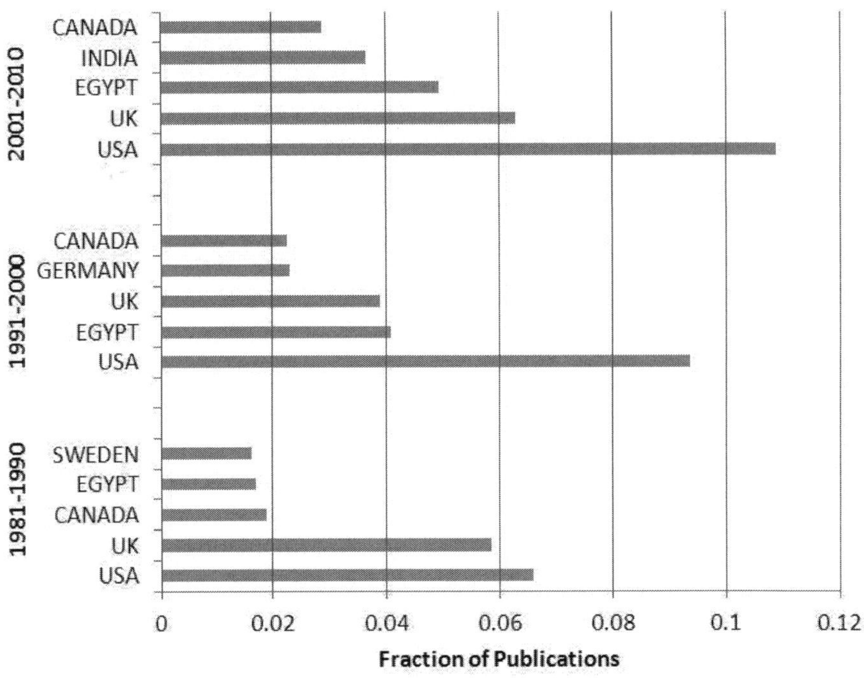

Figure 4a: Top Five Collaborating Partners for Kuwait. US, UK, Canada, and Egypt were consistently among the top five countries with most frequent joint publications with Kuwait and Saudi Arabia during 1981-1990, 1991-2000, and 2001-2010.

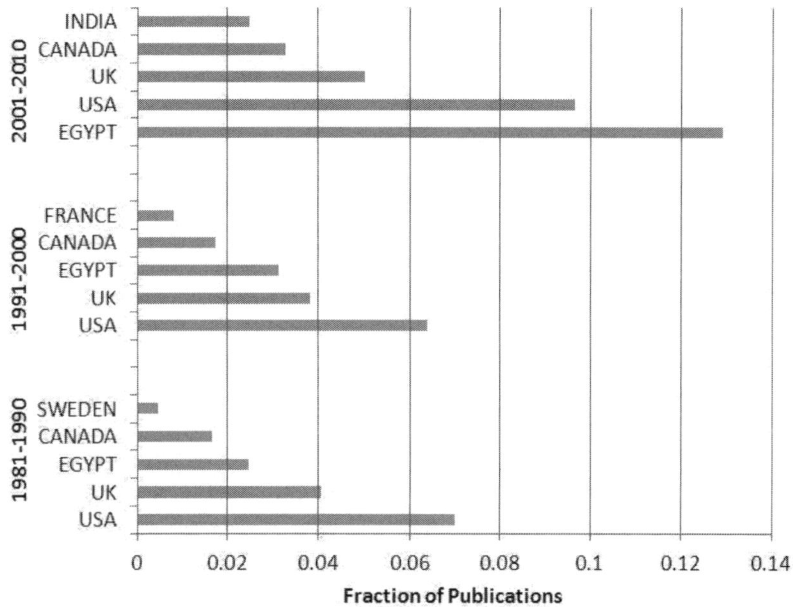

Figure 4b: Top Five Collaborating Partners for Saudi Arabia.

The results convey some expected and some surprising results. The United States (US) and the United Kingdom (UK) are among the top five partners for each of the six GCC countries. As two large countries with a long history of involvement in the region and with productive science and engineering sectors, this is not very surprising. More surprising perhaps is the presence of Canada among the top five partners in almost all countries and time periods. The role of Canada could be explained by similarities in terms of Canada's strong oil and gas and extractive sectors.

Apart from the UK and North America, Egypt is a consistently preferred partner across the GCC countries (with Oman as an exception). This can be at least partly explained by the large number of Egyptian researchers and academics in the GCC, as well as due to education and training links that date back several decades along with cultural and historical ties. An interesting trend is that Egypt has surpassed the US as the largest partner with Saudi Arabia in co-authorships in 2001-2010 period, and the opposite has taken place for Qatar where the US has become the dominant partner replacing Egypt in this period.

We find that GCC countries have also engaged in collaborations with other industrialized countries, such as Germany and France. Oman shows the greatest variety with top collaborating countries including Japan, Australia, and Netherlands in addition to US, UK, and Canada over three decades. In the last time period analyzed (2001-2010), India has emerged as a major partner for Kuwait, Saudi Arabia, and Oman.

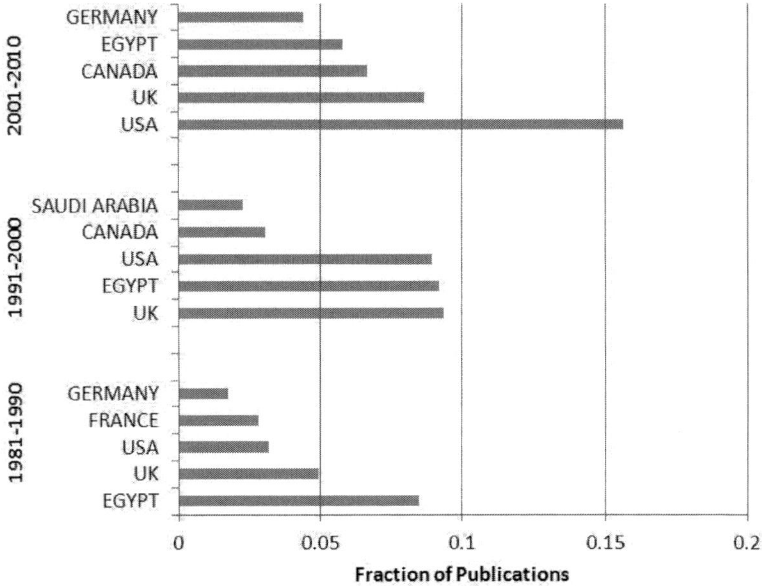

Figure 5a: Top Five Collaborating Partners for UAE.

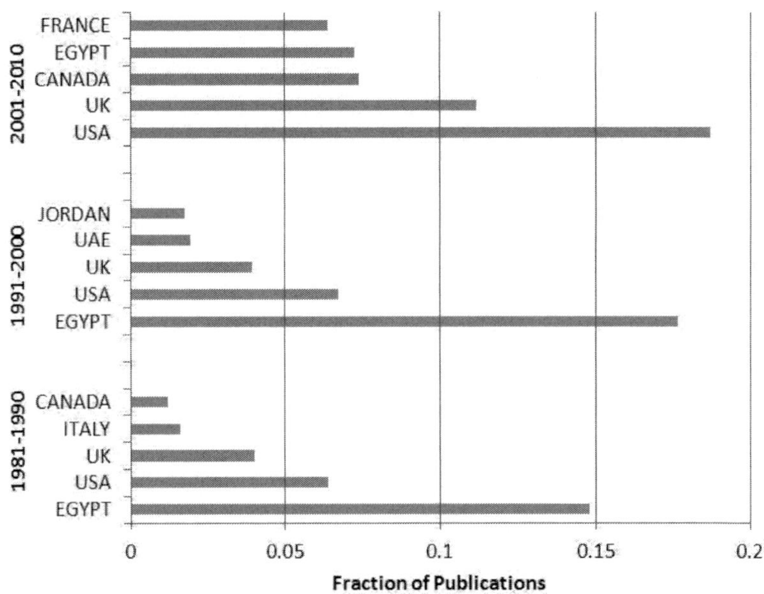

Figure 5b: Top Five Collaborating Partners for Qatar. US, UK, and Egypt are most frequent partners since 1981.

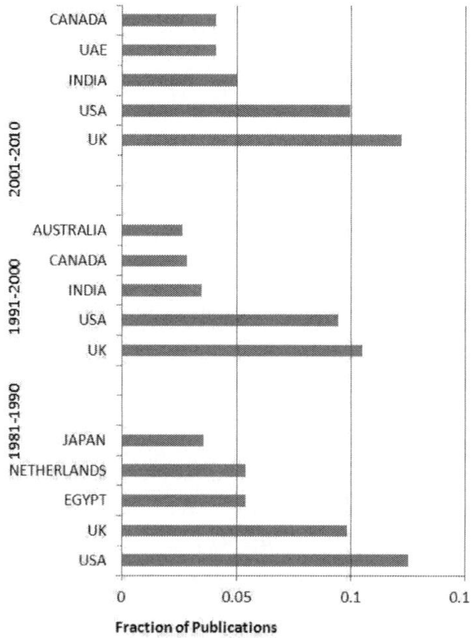

Figure 6a: Top Five Collaborating Partners for Oman.

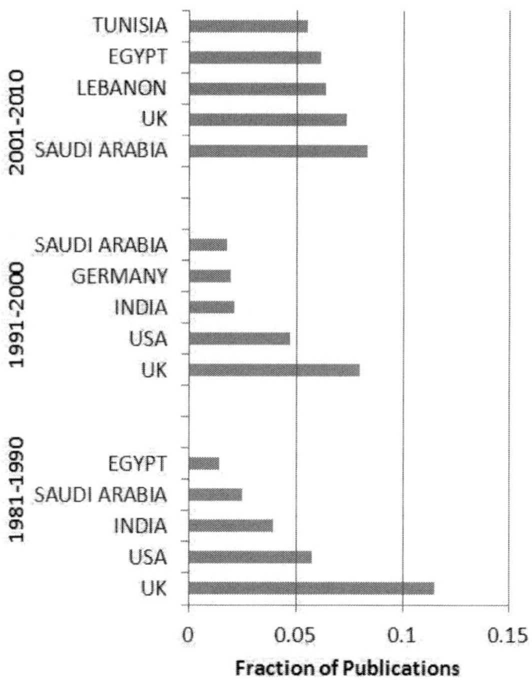

Figure 6b: Top Five Collaborating Partners for Bahrain. Both Oman and Bahrain show greater variety in collaborating partners over time as compared to other GCC states.

Affinity for Research Collaborations

While counting total publications with co-authorship across pairs of countries provides a simple measure for characterizing the geography of collaborations, a better metric is one that controls for a country's size of scientific output. The fact that US and UK appear as top partners in co-authorship for GCC publications is due to the fact that these countries have substantial number of total annual publications. The 'preference' patterns of collaboration can be measured by evaluating how much a country is collaborating as compared to what would be expected given its total publications. We employ a *collaboration affinity*[6] metric based on co-authorship preference developed in (Schubert and Glanzel 2006), wherein we take the ratio of joint publications between two countries normalized by the size (in terms of publications) of the collaborating countries. When the extent of collaborations is balanced, the affinity will have a value of 0. Where there is more collaboration than what would be expected, the affinity value will be greater than zero, and when there is less, it will be less than zero (i.e. negative). When there is no collaboration the affinity is undefined (see appendix for details on calculating the affinity values).

We conducted the affinity analysis for the GCC states, the additional 11 MENA countries, as well as ten selected comparison countries: Turkey, India, UK, Germany, France, Canada, USA, Singapore, South Korea, and China. These comparison countries included common partners of the GCC states (shown in Figures 4-6), as well as three East Asian countries of Singapore, South Korea and China that have undergone major development in their scientific research capacity and expansion.

Figures 7-9 show the values of affinity between each country pair as a heat map for the decades 1981-1990, 1991-2000, and 2001-2010. The dark colored regions indicate low to no collaboration, whereas successively lighter colors indicate regions of intense activity (joint publications). The GCC and other MENA countries are ordered according to geographic contiguity from Kuwait to Morocco in the figures. This ordering allows for easily identifying geographic preference clusters. The comparison countries from Turkey to the US, however, do not have such geographic ordering in the heat maps.

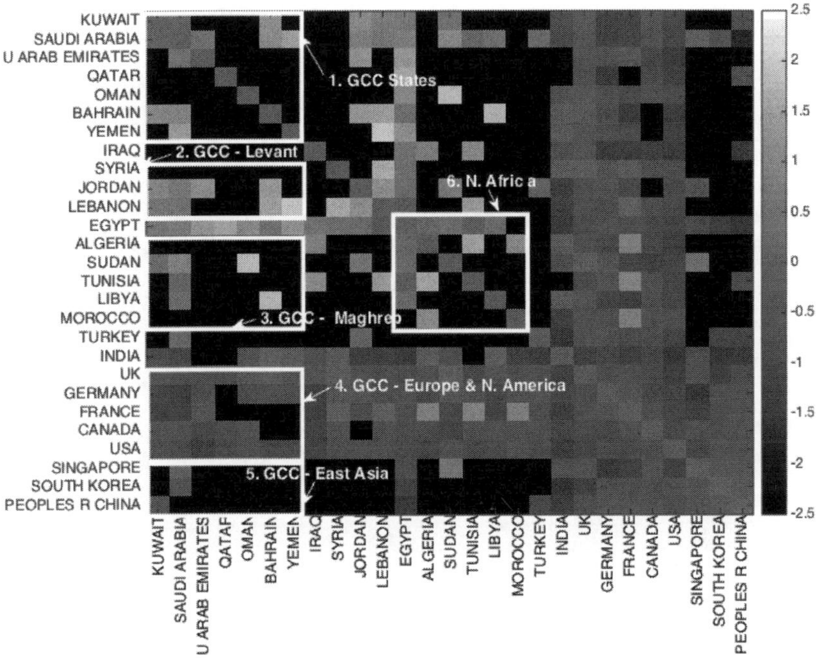

Figure 7: Collaboration affinity of the GCC with itself and other regions between 1981-1990.

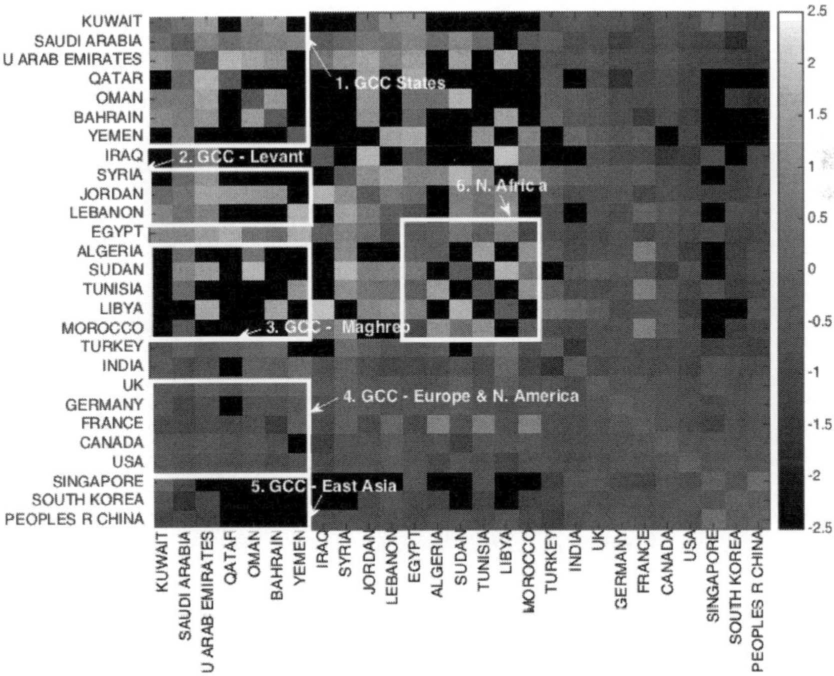

Figure 8: Collaboration affinity of the GCC intensifies in 1991-2000 decade.

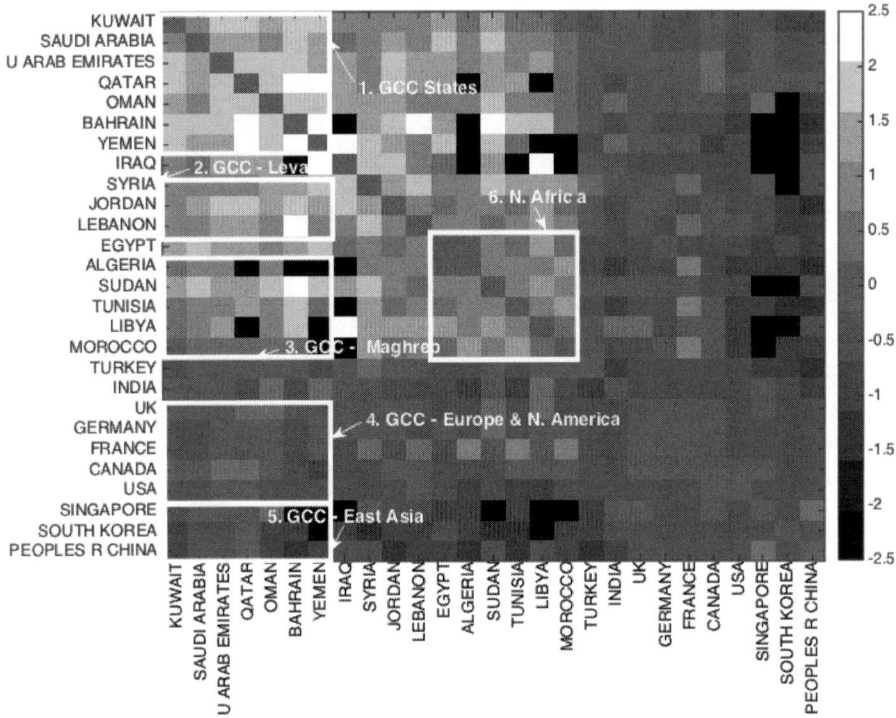

Figure 9: Collaboration affinity of the GCC intensifies in 2001-2010.

A trend of growing affinity among the GCC states is prominent. Starting with almost zero interaction (of co-authorship) in the 1981-1990 decade (Figure 7), we find that collaborations have intensified over each decade (see box 1), with a much greater intensity of collaboration between 2001 and 2010 among GCC (Figure 9). The affinity for other MENA countries in the Levant and Maghreb has also consistently increased (box 2 and 3), however affinity with industrialized countries has risen but remains on the cooler side.

With North African states, Bahrain has comparatively the strongest links (with the exception of Algeria) in the 2001-2010 period. Saudi Arabia shows strong links with Egypt and Sudan (Figure 9). It is interesting to note that the affinity among the North African states (box 6 in Figure 9) is not as strong as the affinity among the GCC states (see box 1 in Figure 9). There are, however, comparatively stronger links between Algeria and Tunisia with Morocco. This pattern of weak and almost non-existent linkages between Maghreb countries in the 1990s was also identified in previous analysis (Zahlan 2012, 88). Algeria, Tunisia, and Morocco show strong links with France, that have persisted across all the three decades (from 1981), although the level of affinity is declining, which is interesting in and of itself.

In explaining collaborations between MENA countries, it has been discussed that the patterns are between those countries that employ substantial Arab expatriate professors and professionals such as Oman, Saudi Arabia, UAE, Kuwait, and Libya, as well as those that provide the bulk of the expatriates: Egypt, Jordan, and increasingly the Maghreb countries. This pattern of mobility has impacted co-authorship links. It has also been discussed that injection of new faculty leads to increased output of the hiring country. However, some of the published work may have been conducted elsewhere in the scientists' previous occupation, and there is thus a need to interpret the data with caution (Zahlan 2012).

For North America (Canada and US), it can be noted that UAE and Qatar have developed a positive affinity level with Canada. The affinity levels with USA for the GCC states remain at or below zero indicating that although in terms of total publications, the US is an important partner for the GCC, when the publication fraction is accounted in terms of total publication activity of the US, there is less collaboration than the expectation.

To assess how affinity for new partners has evolved, we counted the number of countries for which each GCC state had an affinity greater than 0 across each time period. The results are shown in Figure 10 (detailed data with partner names are given in Table 1 in the appendix).

We find that Qatar has had the highest increase, moving from one preferred partner (Egypt) in 1981-1990 period to acquiring eighteen preferred partners in 2000-2010. The list of partners (Table 1) indicates that while most of the countries for which Qatar has developed an affinity are within the MENA region, there are also non-regional countries such as Canada, UK, France and Singapore. UAE shows a similar trend in increasing its partners from three (Egypt, Jordan, and Saudi Arabia) in 1981-1990 to twenty-one in 2001-2010. The new partners include all the regional countries (from Morocco to Kuwait) as well as Canada, Singapore, UK, Turkey and India. Oman also shows mostly similar trends as the UAE.

The results for Saudi Arabia shows that it has not significantly expanded its international linkages (among the set of countries analyzed). Its preferred partners have changed from twelve to seventeen over thirty years, with the new partners being Qatar, Oman, Algeria, Syria and Morocco. There are no non-regional countries in the set of countries for which Saudi Arabia has a positive affinity (Table 1).

Overall, we find that the affinity between GCC countries has grown. Given the literature on the role of proximity, this could be a beneficial development for the formation of regional capacity. There are also increased ties with Maghreb countries (Morocco, Algeria, Tunisia) that had not been previously present. The Maghreb countries were historically under represented in the GCC expatriate population, however, there has been increasing migration of scientists from that region (Zahlan 2012). It is also interesting to note that the smaller GCC countries (Qatar and UAE in particular) have increased ties with non-regional partners, which is not the case for Saudi Arabia or Kuwait. Saudi Arabia has expanded its links with Egypt, and its co-authorship affinity remains focused within Arab countries.

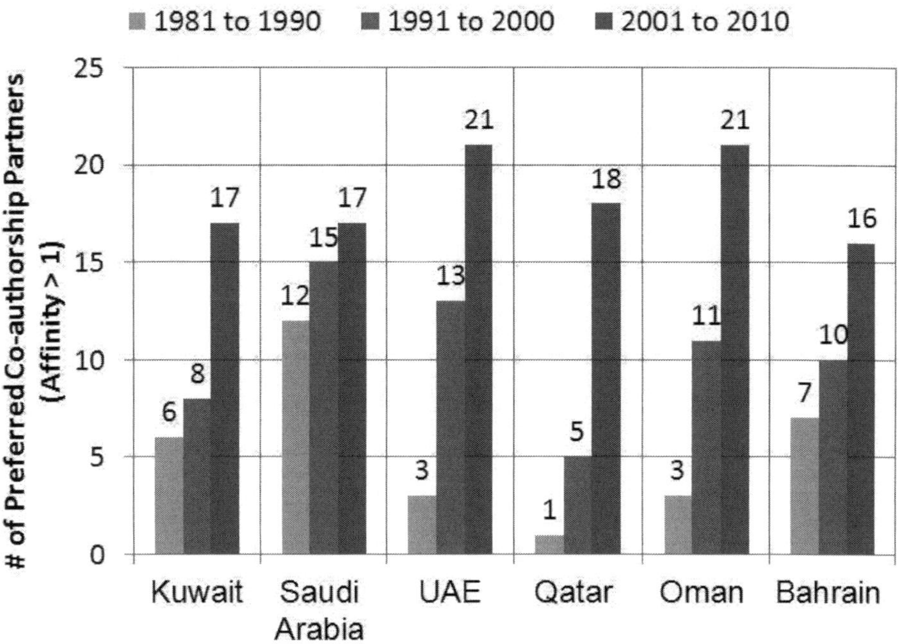

Figure 10: Number of countries with affinity greater than one (indicating preferred co-authorship) for each GCC state over 1981-1990, 1991-2000 and 2001-2010 periods. Table 1 shows names of partner countries.

A key topic we have not explored here is of the quality and impact of the research output. We have focused only on the quantity (annual number of publications) and other metrics based on this measure. For nascent programs and nations seeking to establish a local scientific base, the quantity metrics are important early indicators of progress. The quality, however, will have to be studied since that is of ultimate concern towards efforts of building local capacity that can compete in the global scientific enterprise. Investigating quality across collaboration links can shed light on potential impacts of international research for the GCC. Previous research has found that papers with international co-authors are more cited on average than papers with only domestic authors (Adams 2013; Schubert and Glanzel 2006).

The Impact of Near Ties: Accelerating Local Research in the GCC

International collaborations have been an intrinsic part of scientific research, and cooperation, collaboration, and competition shaped scientific discoveries and progress over the past three centuries. How GCC states engage and embed themselves in this sphere will influence their future development in critical ways.

As compared to 1980s and 1990s, the region has advanced its research output since the turn of this century. As a whole, the GCC countries have sought and extended research linkages, and the ties among the GCC have grown when viewed from a research co-authorship lens. The collaboration patterns, evident from our analysis, can be attributed to geographic and linguistic ties – but there are other factors of researcher mobility and migration also at play. The opportunities in research and teaching in the GCC have led to increasing ties with the Maghreb and the Levant. Saudi Arabia, the largest country, has made the longest strides in increasing output, however it has remained regionally engaged. It is the smaller states of Qatar and UAE that have led in global engagement and have extended their involvement in the international science and technology research realm.

All in all, the increased collaborative links provide indications of a region that is beginning to better integrate, and if such trends continue they can provide crucial momentum in increasing knowledge exchange. However, it would be important to further investigate the types of collaborations that are leading to co-authorship links before a better assessment can be made regarding impact on local capacity development for the long-term. It has been noted earlier that collaboration on the level of individual scientists is important, and has existed across the MENA region. However, organizational level and institutional links have been weak and insufficient for allowing coordinated research efforts to tackle shared regional challenges. The challenges of water and food security, public health, societal development and environmental protection will require cooperative and combined efforts. It is towards addressing these issues that international collaboration within and outside the region can be strategically steered for shared gain and accelerated development.

References

Adams, Jonathan. 2013. "The fourth age of research", *Nature*, 497: 557-560.

Adams, J. (2012). Collaborations: The rise of research networks. *Nature*, 490(7420): 335–336.

Adams, J., Christopher, K., Pendlebury, P., Hook, D., and Wilson, J. (2011), "Exploring the changing landscape of Arabian, Persian, and Turkish research", Global Research Report, Thompson Reuters.

Heitor, Manuel. 2015. "How University Global Partnerships May Facilitate a New Era of International Affairs and Foster Political and Economic Relations", *Technological Forecasting & Social Change*, 95: 276-293.

Al-Ohali, Mohammad and Steve Burdon. 2013. "International Collaboration." In *Higher Education in Saudi Arabia: Achievements, Challenges and Opportunities*, edited by Larry Smith and Abdulrahman Abouammoh, 159-166. Dordrecht: Springer

Andersson, T and Djeflat, A, Eds. 2013. *The Real Issues of the Middle East and the Arab Spring: Innovation, Technology, and Knowledge Management*, New York: Springer. DOI: 10.1007/978-1-4614-5248-5_18.

Bukhari, Fawzy and Brian Denman. 2013. "Student Scholarships in Saudi Arabia: Implications and Opportunities for Overseas Engagement." In *Higher Education in Saudi Arabia: Achievements, Challenges and Opportunities*, edited by Larry Smith and Abdulrahman Abouammoh, 151-158. Dordrecht: Springer

Brinkely, Ian, Will Hutton, Philippe Schneider and Kristian Coates Ulrichsen. 2012. *Kuwait and the Knowledge Economy*. Report by The Work Foundation and the Kuwait Programme on Development, Governance and Globalisation in the Gulf States.

CCE. 2015. Accessed: Nov 30. http://www.cce.edu.om/default.aspx

Commins, David. 2012. *The Gulf States: A Modern History*. New York: I. B. Tauris & Co.

CNRE. 2015. Accessed: Nov 30. https://cnre.mit.edu.

DoS. 2015. Remote sensing cooperation agreement between United States of America and Saudi Arabia. Accessed: December 11 http://www.state.gov/documents/organization/211211.pdf

Glanzel, W. 2001. "National characteristics in international scientific co-authorship relations", *Scientometrics*, 51(1): 69-115.

Al-Sultan, K. S., and Iyad T. Alzaharnah. 2012. "Academia-Industry Innovation Linkages in the Case of Saudi Arabia: Developing a University-Industry Triple-Helix Framework to Promote Research and Development Collaboration". In *Global Innovation Index 2012*, edited by Soumitra Dutta, 89-95.

France: INSEAD.

Godin, B., and Ippersiel, M. P. 1996. "Scientific collaboration at the regional level: the case of a small country", *Scientometrics*, 36(1): 59-68.

GUTech. 2015. German University of Technology Oman. Accessed Nov 30. http://www.gutech.edu.om

ODO. 2015. "Science & Tech City projects get fillip with setting up of panels". *Oman Daily Observer*, May 12. http://omanobserver.om/science-tech-city-projects-get-fillip-with-setting-up-of-panels/

Hajjar, D. P., G. W. Moran, Afreen Siddiqi, J. Richardson, Laura D. Anadon, and Venkatesh Narayanamurti. 2014. "Prospects for Policy Advances in Science and Technology in the Gulf Arab States: The Role for International Partnerships", *International Journal on Higher Education*, 3(3): 45-57.

HKS. 2015. Accessed: Nov 30. http://belfercenter.hks.harvard.edu/project/64/middle_east_initiative.html?page_id=422.

IMCO. 2015. Accessed: Nov 30. http://www.imcoman.net/glance.php

Katz, J. S., and Martin, B. R., 1997. "What is research collaboration?", *Research Policy*, 26:1-18.

KACST. 2015. Accessed: December 11. http://www.kacst.edu.sa/en/about/media/news/Pages/news185.aspx

KFAS. 2012. KFAS Strategic Plan 2012-2016. http://www.kfas.org/strategic.html

KISR. 2015. Accessed: Nov 30. http://www.kisr.edu.kw/en/about-us/history.

KU. 2015. Accessed: Nov 30. http://www.kuniv.edu/ku/AboutKU/StrategicPlan/Mission/index.htm.

http://www.jadwa.com/en/download/2014-budget/saudi-arabias-2014-budget

Landini, Fabio, Franco Malbera and Roberto Mavilia. 2015. "The structure and dynamics of networks of scientific collaborations in Northern Africa". *Scientometrics*, 105:1787-1807.

Ma, A., Raul Mondragon, and Vito Latora. 2015. "Anatomy of funded research in science". *Proceedings of the National Academy of Sciences,* 112(48):14760-14765.

Smith, Larry and Abdulrahman Abouammoh, Eds. 2013. *Higher Education in Saudi Arabia: Achievements, Challenges and Opportunities*, Dordrecht: Springer

Lippman, Thomas W. 2012. *Saudi Arabia On The Edge*. Washington D.C: Potomac Books.

LSE. 2016. Accessed: January 13. http://www.lse.ac.uk/middleEastCentre/kuwait/opportunities/Opportunities.aspx

May, R. 1997. "Scientific Wealth of Nations", *Science*, XX.

Newman, M.E.J., 2001. "The structure of scientific collaboration networks", *Proceedings of the National Academy of Science*, 98 (2): 404-409.

NYIT. 2015. Accessed: December 11. http://www.nyit.edu/bahrain/

Reiche, D. 2010. "Renewable energy policies in the Gulf countries: A case study of the carbon-neutral "Masdar City" in Abu Dhabi. *Energy Policy*, 38: 378-382.

QSTP. 2015. Accessed: December 11. http://www.qstp.org.qa/home/free-zone

Sorenson, O., J. W. Rivkin, L. Fleming, Complexity, networks and knowledge flow. *Research Policy* **35**, 994–1017 (2006).

Schubert, Andras and Wolfgang Glanzel. 2006. "Cross-national preference in co-authorship, references, and citations". *Scientometrics*, 69 (2) 409:428.

TAMU. 2015. Accessed: December 11. https://engineering.tamu.edu/news/2015/10/26/texas-am-at-qatar-launches-the-oryx-gtl---gas-to-liquid-excellence-program

Trencher, Gregory., Yarime Masaru, Kes B. McCormick, Christopher N. H. Doll, and Steven B. Kraines. 2014. "Beyond the Third Mission: Exploring the Emerging University Function of Co-Creation for Sustainability", *Science and Public Policy*, 41:151-179.

TRC. 2015. Accessed: Nov 30. https://home.trc.gov.om/tabid/40/language/en-US/Default.aspx

TWAS. 2015. "Oman: Advancing Science, advancing prosperity". *The World Academy of Sciences*. February 6. http://twas.org/article/oman-advancing-science-advancing-prosperity

UAEE. 2015. United Arab Emirates Embassy. http://www.uae-embassy.org/uae-us-relations/partnerships#academic

UML. 2015. Accessed: Nov 30. http://www.uml.edu/News/press-releases/2014/UMLKuwait121014.aspx

WoS. 2015. Web of Science Core Collection. Accessed: December 14. http://wokinfo.com/products_tools/multidisciplinary/webofscience/

Wuchty, S., Jones, B. F., Uzzi, Brian. 2007. "The Increasing Dominance of Teams in Production of Knowledge", *Science*, **316**, 1036-1039.

Zahlan, Antoine B. 2012. *Science, Development, and Sovereignty in the Arab World*. New York: Palgrave Macmillan

Notes

1 King Abdullah University of Science and Technology, in Saudi Arabia, was set up with an aim of attaining a global ranking in the top twenty science and technology universities within ten years (Al-Ohali and Burdon, 2013). Masdar City in UAE, intended to be a hub for clean technology innovation, was created with a vision to be the world's first "carbon-neutral city" (Reiche, 2010).

2 We used the whole counting method. If a paper had two authors, one with an address in Bahrain and one in UAE, then the paper would be counted once for Bahrain papers and once again for UAE papers.

3 For an accurate measure, the total number of researchers in the country should be used to compute productivity. However, the total population can be used as a proxy when data of number of researchers is not available (May, 1997).

4 The global share included adjustment for full counting of papers with co-authors from multiple countries. We only accounted for English articles (which may only partially represent publications of Iran and Turkey that produce many scientific research publications in Farsi and Turkish respectively).

5 We counted a publication with an author with address in country *i* and an author with address in country *j* as a joint publication between countries *i* and *j*. We denote total joint publications between *i* and *j* in a time period as x_{ij}, and for a country *i*, with total X_i publications in that time period. We then compute the fraction of joint publications as x_{ij}/X_i.

6 **Collaboration Affinity Measure:** We quantify the number of joint publications between pairs of countries using address information of authors on publications. A publication in which at least one author has an address in country i, and another author has an address in country j, is counted towards the collaborative publications total for the pair country i and j.

We define collaboration affinity, α_{ij}, of country i and country j based on the co-authorship preference measure developed in (Schubert and Glanzel, 2006) as:

$$a_{ij} = \log_{10}\left[\frac{x_{ij} / X_i}{X_i / (X_{word} - X_i}\right] \quad (4)$$

where xij is the number of joint papers between country i and country j, Xi is total number of publications of country i, Xj is total number of publications for country j and Xworld is total world publications in the given time duration of analysis.

The numerator in the equation is the fraction of publications of country i in which authors from country i and country j collaborated. The denominator is the fraction of country j's publication within overall publications (in a world in which country i is not accounted) – it provides a measure of the size of country j's output within global scientific output. The α_{ij} is a ratio that provides a relative measure of engagement between i and j in which the size of collaborating country (j) for a given country i is adjusted.

Note, that we use the log function to scale the alpha parameter (which is not done in the Schubert and Glanzel metric). Another important difference in our case from Schubert and Glanzel's measure is that we use total publications data (and not only international collaborations data for each country) in the ratio, thus domestic papers (in which all authors are in one country) also get included in the total publications count.

When the fraction of joint publications between *i* and *j* equals the fractional size of *j* within total scientific output, the extent of collaboration will be balanced, and the fraction will be 1, and α_{ij} will have a value of 0. Where there is more collaboration than what would be expected given the size of *j*, α_{ij} will be greater than zero, and when there is less, α_{ij} will be less than zero (i.e. negative). When there is no collaboration between *i* and *j*, α_{ij} is undefined.

A note of caution is that in this study we have not distinguished between single and multiple affiliation of authors. Thus, if an author has a joint affiliation in institutions in two different countries the paper will get counted as a joint paper between the countries. The collaboration in this case would therefore not be inter-individual, rather inter-institutional through a joint researcher (Katz and Martin, 1997).

Table 1: List of countries with affinity greater than one for each GCC state

Kuwait																					
1981-1990	BHR	JOR	LBN	EGY	SDN	SAU															
1991-2000	UAE	BHR	EGY	LBN	OMN	SAU	JOR	SGP													
2001-2010	BHR	YEM	QAT	UAE	OMN	SDN	EGY	SAU	SYR	JOR	LBN	IRQ	LBY	TUN	ALG	MOR	IND				
Saudi Arabia																					
1981-1990	YEM	BHR	SDN	EGY	JOR	LBN	LBY	UAE	TUR	TUN	KWT	SGP									
1991-2000	SDN	UAE	YEM	EGY	BHR	OMN	JOR	LBN	ALG	QAT	TUN	KWT	SYR	IRQ	TUR						
2001-2010	BHR	SDN	EGY	YEM	QAT	UAE	JOR	LBY	KWT	OMN	ALG	LBN	TUN	SYR	IRQ	MOR	TUR				
UAE																					
1981-1990	EGY	JOR	SAU																		
1991-2000	QAT	LBY	JOR	OMN	SDN	EGY	KWT	BHR	SYR	SAU	LBN	IRQ	UK								
2001-2010	QAT	OMN	BHR	JOR	KWT	LBY	YEM	IRQ	SDN	SAU	EGY	LBN	SYR	ALG	TUN	MOR	CAN	SGP	UK	TUR	IND
Qatar																					
1981-1990	EGY																				
1991-2000	UAE	EGY	JOR	SAU	TUR																
2001-2010	YEM	BHR	UAE	OMN	JOR	KWT	SDN	IRQ	EGY	SAU	LBN	TUN	SYR	MOR	CAN	UK	FRA	SGP			
Oman																					
1981-1990	SDN	EGY	UK																		
1991-2000	SDN	UAE	BHR	JOR	KWT	SAU	EGY	ALG	TUR	IND	UK										
2001-2010	YEM	UAE	SDN	BHR	QAT	JOR	KWT	SYR	IRQ	ALG	SAU	EGY	LBN	LBY	TUN	MOR	IND	UK	SGP	CAN	TUR
Bahrain																					
1981-1990	LBY	LBN	JOR	SAU	KWT	EGY	IND														
1991-2000	LBY	OMN	UAE	JOR	KWT	SAU	EGY	TUR	IND	UK											
2001-2010	SDN	LBN	QAT	YEM	UAE	OMN	LBY	KWT	SAU	TUN	JOR	SYR	EGY	MOR	TUR	UK					

3

Science, Technology and Innovation in the GCC: Labour Market Challenges

Aruba Khalid

Introduction

Over the past decade policymakers across the GCC have widely acknowledged the importance of developing strong science, technology and innovation (STI) sectors. The development of STI sectors is viewed not only as an important feature of region-wide economic diversification initiatives, to shift the GCC away from concentration on national hydrocarbon sectors, but also as a critical aspect of achieving long term economic sustainability and contributing to global intellectual and technological advancement in a post oil stage (Bahar 2013). In order to achieve these aspirations and raise the significance of STI sectors in their national economies, policymakers in the region have designed various developmental plans and strategies. The states plans are notably varied; key recurring feature amongst each plan is the notable emphasis placed on increasing human capital, particularly in science and technology. This has led to region wide development of educational institutions, training facilities and academic cities.

Regional initiatives coincide with an ongoing global trend of strengthening education in STI fields. Advanced and emerging economies, such as the UK, US, India and China, have been focusing specifically on the significance of their STI workforces having recognized that economic competitiveness in the future will be founded chiefly on knowledge industries. These trends indicate that economies will increasingly compete with each other for top talent in STI fields thus paving the way for a 'global skills race' (Brown and Ashton 2008, 4). This poses a serious challenge for the GCC due to certain characteristics of traditional labor markets in the region.

Historically, GCC economies have relied heavily on a somewhat transient external labor force that has been bound by temporary contracts and typified by constant churn and low skill levels. It is found that workers with tertiary education account for less than 25 per cent of the GCC's labor market, comprising of both a majority of lower skilled workers amongst foreign and national workers. Consequently, developing a highly skilled

STI workforce in the region will require drawing and retaining highly skilled professionals to educate, train and develop local talent and build the GCCs nascent knowledge industries in the near term and strengthen labor force competitiveness in the long term. However, due to labor market particularities in the context of foreign labor i.e. temporary contracts and weak residency and employment security highly skilled professionals may be lost to other more attractive economies. This poses significant challenges to the development of a highly skilled STI workforce in the GCC.

The purpose of the present study will be to assess these challenges and consider politically and culturally acceptable models to address them. The study will primarily assess the potential for a global skills race by surveying regional and international initiatives to develop national STI talent. It will then attempt to determine possible challenges the GCC faces in building an STI workforce by considering key features of regional labor markets. Lastly, the study will consider how other economies have dealt with and are dealing with similar challenges by surveying parallel mechanisms that facilitate international recruitment, and will explore basic recommendations for the GCC.

The Establishment of STI Sectors in the GCC

Over the past two decades, GCC policymakers have been aspiring to develop national science, technology and innovation sectors, along with other key industrial sectors, in order to diversify their economies. In more recent times, these aspirations have started to materialize through the establishment of strategic plans such as Saudi Arabia's National Science, Technology and Innovation Plan (NSTIP), Abu Dhabi's Science, Technology and Innovation (STI) Vision and Qatar's National Research Strategy. These plans outline each states vision for the development of their STI sectors or call for the establishment of special bodies and committees specifically targeted at the development of STI industries and the implementation of concrete strategies. For instance, Saudi Arabia's NSTIP outlines the Kingdoms long term vision to build and develop a knowledge based economy by establishing a range of programs targeted at the development of key strategic industries such as water technology, biotechnology, advanced materials and nanotechnology. The NSTIP also highlights the development of the country's human resource capacities, strengthening research capacity and enhancing the legislative and institutional framework as some of its key goals (Alsodais 2013). Similarly, Abu Dhabi's Science, Technology and Innovation (STI) Vision seeks to increase the development of certain STI industries such as semiconductors, clean technology and information and communications technology. It also establishes the Abu Dhabi Technology Development Committee (TDC) to support the development of the emirates STI sector and act as a forum for government, business and academia to interact and collaborate, in line with the Abu Dhabi Policy Agenda 2007-2008 and Economic Vision 2030 (Abu Dhabi TDC 2014). Qatar's National Research Strategy outlines a more holistic approach to enhance the states innovation capabilities by focusing on the importance of strengthening research capacity in its five priority themes which include enterprise, energy and environment, computing and information technologies, health and social sciences, arts and humanities. The Strategy follows the mandate of Qatar's National Vision 2030 and establishes key goals and objectives under each pillar all focused on enhancing Qatar's innovation potential, increasing diversification and achieving a sustainable economy and society. Kuwait on the other hand has not established a specific plan targeted at the development of STI sectors alone. However, the Kuwait Foundation for the Advancement of Sciences, the Kuwait Institute for Scientific Research and Kuwait University have launched various programs together with the London School of Economics assessing the importance of developing the national knowledge economy in Kuwait indicating the increasing acknowledgement of the importance of developing STI sectors in the state (Brinkely 2012).

Increasing Emphasis on STI Skills in the GCC Through Expansion of Educational Facilities

Such plans and strategies indicate that the Gulf States have adopted somewhat varied approaches to building their STI sectors and are focusing on different areas of science, technology and innovation. Whereas Saudi Arabia and Abu Dhabi have formulated particular STI plans targeted at specific sectors and are focusing on the building of institutional and legislative frameworks around them, Qatar has chosen a more rounded approach which seeks to address its overall research and innovation capacity by incorporating health issues and social sciences, arts and humanities into the overall vision. However, one of the key recurring elements in each states STI vision is the notable emphasis placed on the development and accumulation of human capital. Along with addressing other key areas in the development of knowledge and information sectors such as seeking to enhance legislative and institutional framework relevant to STI sectors and strengthening the links between enterprise and business and research and innovation, each states STI vision includes mechanisms and initiatives to enhance education, academic research and training indicating regional acknowledgement of the centrality of human capital to competitive STI industries.

Presently, policy level initiatives, particularly those focusing on human capital development, have started to translate into visible developments. These include large universities, such as the King Abdullah University of Science and Technology in Saudi Arabia and Khalifa University of Science Technology and Research in Abu Dhabi as well as large education and knowledge cities, such as Qatar's Education City and the Sabah al Salem City in Kuwait, which host regional branches of renowned international universities. Furthermore, in order to kick start university-industry collaboration and pave the way for future STI ecosystems, several public private partnerships involving renowned international bodies have also been launched. For example, Saudi Arabia's King Abdullah University of Science and Technology has formed a partnership with Pall Corporation, a global leader in filtration, separation and purification technologies, to collaborate on research focused on water treatment, alternative fuels, and other areas of critical importance. Similarly, Kuwait University and IMEX, world-leading nanotechnology research center of Belgium have formed a long term partnership and collaboration agreement for research and development of innovative silicon solar cells and ABB and Qatar University have started a new industry-academia collaboration by way of a series of student projects entitled, ABB & Qatar University Student Project on Robotics.

The ongoing surge of such initiatives reveals that GCC policymakers have not only acknowledged the importance of developing human capital for strong STI industries in the future by formulating national visions and plans, but have also actively started to implement initiatives targeted at achieving their end. The establishment of a vast number of education, research and training institutions in particular indicates that developing human capital, arguably one of the most critical components of competitive knowledge sectors, is a key regional priority. It would appear therefore that having addressed the major criteria requisite for the establishment of strong STI industries, with particular consideration for the development of human capital through education and training, GCC economies are on a strong path towards building competitive STI industries within the foreseeable future. However, a consideration of similar global trends in this area would indicate that a true assessment of the potential success of the GCC's efforts involves additional challenges.

Global Race for STI Skills

Along with the GCC, advanced economies, such as the United States and the United Kingdom, are also actively strengthening their national STI policies with a particular focus on the significance on developing human capital and beefing up their STI workforces. In 2009, the "Educate to Innovate" campaign was introduced by President Barack Obama in the United States as a national effort calling for representatives from industry, academia, foundations, and nonprofit organizations to collaborate in enhancing the quality of science and mathematics education within the next 10 years. This initiative comes on the back of the US's visibly increased understanding, over the past 2 decades, of the importance of strengthening U.S. science, technology engineering and maths (STEM) concepts and skills particularly in the pre-college education curriculum. Public private partnerships aimed at building the future workforces STEM skills presently account for over USD700 million and have achieved a range of initiatives such as training 100,000 new STEM teachers, strengthening federal investment in STEM and increasing student and worker participation in STEM subjects and fields (Educate to Innovate 2014).

Similarly, the UK Government highlighted the importance of education, particularly in STEM subjects, in its 2011 report, The Plan for Growth, by referring to the development of STEM human capital as the 'foundation of future economic success'. The report emphasized the immense significance of STEM in driving innovation, growth and economic recovery in the future and stressed that it will not only be critical for building STI industries but also to provide people with the scientific literacy and numerical understanding needed for informed debate and decision-making. The report identified the need to increase and widen participation in STEM as well as maintaining and improving the quality of STEM education and ensuring that education is directly suited to industry and society requirements.

Along with advanced economies, key emerging economies such as China, India and Brazil have simultaneously acknowledged the importance of science, technology and innovation for ensuring economic competitiveness in the future and have been actively installing policies and initiatives to build their STI workforces. China has launched a long-term funding program for science and technology which seeks to fuel knowledge-based sectors including investments from the Ministry of Science and Technology totaling approximately USD 32 billion. Results of this funding have already become visible as in 2011 41 per cent of all degrees awarded by Chinese institutions were in a STEM subject, almost twice the proportion of STEM degrees awarded in the UK and three times the rate in the US (Katsomitros 2013). In India, the national Science, Technology and Innovation Policy of 2013 declared 2010 – 2020 the 'Decade of Innovation' and has emphasized the importance of accelerating national innovation efforts and fuelling private initiatives towards this end. For instance, telecommunications giant Intel India launched the National Science, Technology, Engineering, and Mathematics Acceleration Program to strengthen STEM education in India in February 2014. The program focuses on initiatives seeking to enhance creativity and innovation amongst students in India. Similarly, Brazil has also been increasing enrollments in STEM subjects as local companies operating in STI sectors have started to grow and require larger volumes of skilled STI workers. It is estimated that by 2015, Brazil will increase its engineering graduates by 68 per cent and will produce more PhD engineers than the US by 2016 (Isalkar 2014). These trends reveal the onset of a global 'skills race' within which education, knowledge and skills, as embodied in highly talented and skilled individuals, will be the most highly sought after resources. Furthermore, it is assumed that this 'race' will increasingly accelerate in the future as STI industries in emerging economies develop and strengthen and begin to compete more actively with those in advanced economies (Brown and Ashton 2008, 4).

The previous discussion illustrates the increasing significance of STI education globally. However, it is important to note that how effectively a country can accumulate human capital does not solely rely on its education and training capabilities. The rapidly increasing international mobility of highly skilled workers indicates that talent is not necessarily bound to its original location but increasingly flows to economies perceived to be more attractive (Industry Canada 2011). Furthermore, increasing educational facilities may not be able to address gaps between national human capital development and national STI industries labor requirement due to volume shortages or specific skill shortages particularly in the short or medium run. Additionally, education and training facilities themselves will require vast number of highly skilled professionals which may not be available through existing educational institutions. Consequently, a country's competitiveness in supplying its national STI sectors with the required highly skilled workers will rely on the sophistication and effectiveness of its national labor markets which must supply national STI industries with the requisite workforce. Labor market policies and particularities must therefore also be assessed in order to determine the potential of a country's human capital accumulation efforts.

Key Labor Market Characteristics in the GCC

In order to determine the GCC's strength in building its STI industries and supplying them with the requisite human capital it will be crucial to understand existing labor markets in the region and their key defining characteristics. GCC economies, and their corresponding labor markets, have notable variances amongst them and cannot entirely be unified under a singular model. However, certain similarities exist across them and become apparent when their labor markets characteristics are assessed. These characteristics include the particularly large number of foreign workers, both relatively and absolutely, the institutional structures, such as the sponsorship system, through which these foreign workers are employed, and consistently low regional labor productivity. Although there are various other labor market particularities that are common amongst the GCC economies, such as an overwhelming large presence of male workers relative to female workers, for the present discussion we will focus on the previous three characteristics.

Large Populations of Expatriate Workers

Firstly, one of the key characteristics of GCC labor markets is the large population of expatriate workers. According to United Nations Population Division data, international migrants accounted for 84 per cent of the UAE's population in 2013, allowing it to rank the highest in terms of international migrant stock. In the same year Qatar ranked third, Kuwait seventh, Bahrain thirteenth, Saudi thirty-third and Oman thirty fourth. The UAE, Qatar, Kuwait and Bahrain actually maintain a larger percent of international migrants relative to national citizens (Figure 1). When absolute figures are considered, Saudi Arabia appears to have the largest number of migrants, reaching almost 9.1 million in 2013, followed by the UAE at 7.8 million (Figure 2). These figures put the Arab Gulf as the third largest receiving region for global migrants, after North American and the European Union (Lori 2012, 3).

Labour inflows into the GCC come from a wide range of countries. Although exact figures of the expatriate population broken down into separate nationality groups are not regularly recorded by authorities, common population counts from commercial industry sources find that the most significant shares of the expatriate population are accounted for by South and South East Asians, Arabs and to a smaller degree, by Europeans. Workers from India, Pakistan, Egypt, Jordan, Bangladesh, the Philippines, Sri Lanka, Nepal, Syria and Sudan constitute the 10 largest expatriate population groups in the UAE, Kuwait, Qatar and Saudi Arabia according to Business Qatar magazine. Large numbers of workers from the UK are also commonly noted to form a significant share of GCC expatriate populations by media sources.

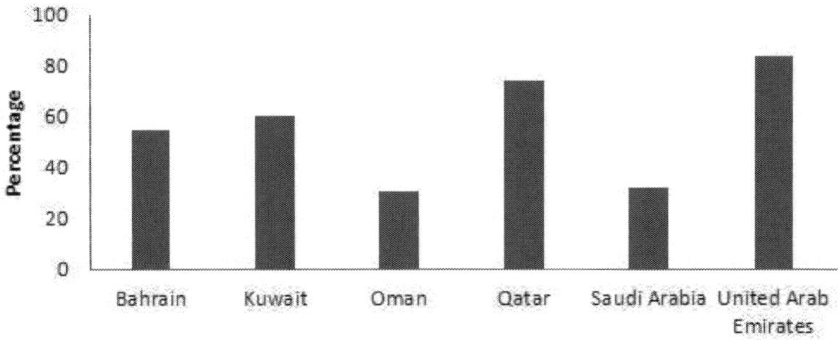

Figure 1: International Migrant Stock as of Share of Total Population in 2013. Data Source: UN Population Division.

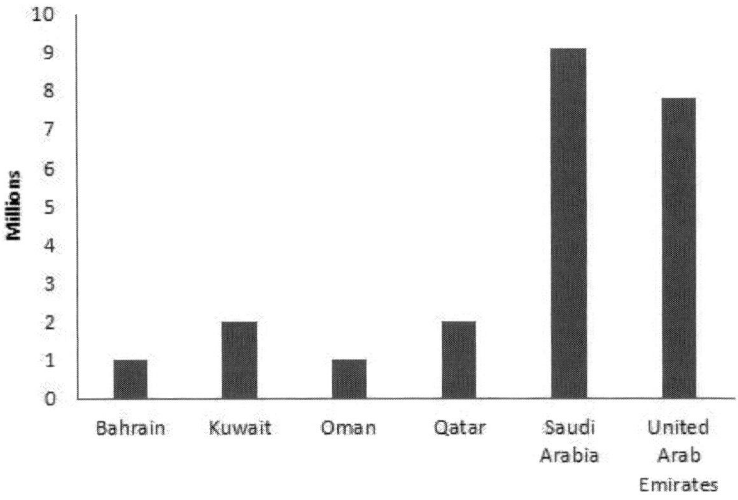

Figure 2: Total International Migrant Stock in 2013. Data Source: UN Population Division.

The Gulf's dependency on foreign workers is a result of several factors, including low labor force participation, especially of women; a small native population; initially low literacy and educational levels; and high non-labor income for nationals. It is important to note however that 'unemployment' has not translated into immediate and visible social consequences given the vast economic resources and continued ability of states to provide public sector employment (Pradhan 2010). Although other economies in the world also host sizeable numbers of foreign workers, such as the EU and the US, large populations of foreign workers in the GCC are bound by an employment system which notably differs from other regions and is unique to the GCC and certain other countries in the Middle East.

Temporary Employment Contracts

The second significant characteristic of GCC labor markets is the institutional framework through which foreign workers are hired and retained in the region. Effectively, foreign workers in the GCC are employed as 'guest workers' through fixed-term employment contracts which must be renewed annually or every two or three years. Unlike the permanent residence and citizenship models of other regions, foreign workers in the GCC are obliged to leave upon the termination of their contract and are thus not officially categorized as 'migrants' but as 'temporary contractual laborers'. The system through which temporary workers are employed is referred to as the Kafala or sponsorship system. This mechanism makes an individual national citizen or company sponsor (known as the Kafeel) legally and economically responsible for the foreign worker for the duration of the contract period and enables Gulf governments to maintain some formal/legal control over how long individual workers can reside in the country (Lori 2012, 4).

In recent times many of the GCC states have discussed the introduction of changes to the sponsorship system but no significant changes have taken place. In 2009, Bahrain announced plans to end the sponsorship system but later withdrew these plans. However, migrant workers in Bahrain are now permitted to change their employers without the consent of their current employers, which was not the case before, and migrants with expiring contracts are given three months to find new employers. Similarly, although Kuwait also announced plans to end the sponsorship system in February 2011, it revoked these plans and introduced other softer changes such as making it easier for migrants to change sponsors and employers. In the UAE, the sponsorship system also continues to remain intact. However, a wage protection system requiring employers to pay the wages of migrant workers directly into bank accounts has been implemented recently. The wage protection system was installed to make it easier to resolve claims of non-payment of wages (Migration News 2012). In Qatar and Saudi Arabia, however, no significant changes have been made to the sponsorship system.

Low Labor Productivity

Thirdly, one of the key characteristics of GCC labor markets is that they have traditionally been characterized by low labor productivity. This is essentially because national oil and gas sectors, which constitute the largest economic sectors in most of the GCC countries, account for only small levels of employment whereas less 'economically significant' sectors such as construction and utilities, government services and other services employ much larger numbers of labor. This has resulted in creating highly uneven employment and labor productivity (Abouchakra 2008, 4). According to the World Bank, oil rents accounted for around 44 per cent of Saudi Arabia's GDP, 22 per cent of the UAE's GDP and 24 per cent Qatar GDP in 2011-2015. In 2008, the oil and gas sector accounted for 47 per cent of GCC countries' GDP, but employed only 1 per cent of the overall population

(Abouchakra 2008, 6). On the other hand, government services accounted for around 20 per cent of total employment in the GCC. These figures indicate that a vast majority of the region's labor force has traditionally been working in sectors which do not contribute significantly to the GDP or to growth. Consequently, labor and even capital productivities in non-oil sectors lag far behind those in oil and gas. For example, in 2005, the GDP labor productivity in GCC countries was USD1.6 million per employee for the oil and gas sector but only USD9, 300 per employee for construction. In the same year, GDP-to-credit capital productivity was USD121 million per unit of credit for the oil and gas sector but only USD1.2 million for construction (Strategy 2008). Such low levels of labor productivity are compounded by the notably low skill levels which characterize the vast majority of the labor force and make the ratio of low skilled labor over medium to highly skilled labor exceptionally high (Hertog 2012). In Qatar, Kuwait, Saudi Arabia, Bahrain and Oman foreign workers possessing some form of tertiary education constituted only approximately 15 per cent of the total population on average in 2008 (IMF 2013, 9). More recent figures indicate that these numbers have likely increased but continue to remain comparatively low. As visible in Figure 3 below, latest available data indicates that the UAE and Saudi Arabia maintain the highest shares of workers with higher education at 25 to 27 per cent respectively followed by Qatar at 21 per cent. The comparatively high rate of workers with tertiary education in Saudi Arabia and Qatar may likely be accounted for by the recent establishment of several academic institutions which would require faculty. The larger share of high-skilled foreign workers in the UAE is largely attributed to its role as a financial and logistics center requiring highly educated workers (IMF 2013, 9). However, overall this reveals that approximately more than 70 per cent of the total population in GCC states do not have higher education and also reveals that not only is labor productivity low because of the relative strength of the natural resource sector but also because the non-oil sector is typified by low skilled labor[1].

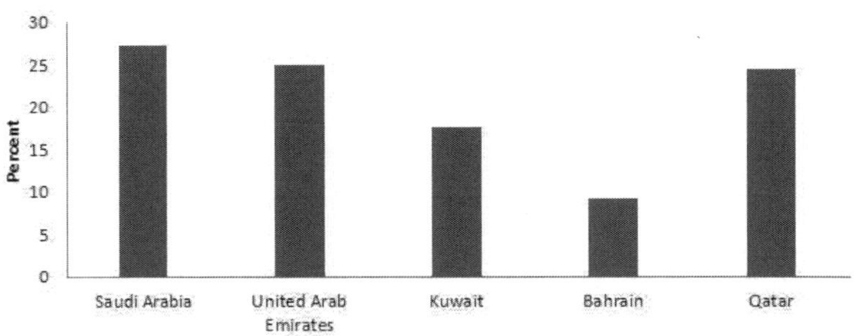

Figure 3: Population with Higher Education in GCC.

As visible in Figures 4,5 and 6 below, the largest share of employed workers in Saudi Arabia, the UAE and Kuwait are involved in occupations of services / service workers, shops and market sales workers and manufacturing respectively.

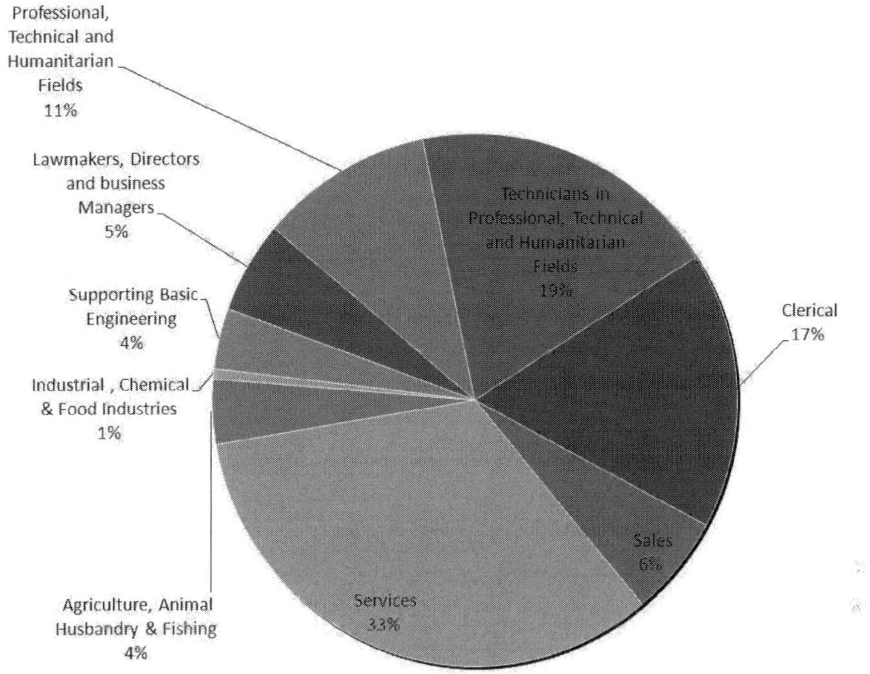

Figure 4: Share of Main Occupation Groups in Employed Population (15 years and above) in Saudi Arabia. Data Source: Labour Force Survey 2014 Round 2, Central Department of Statistics and Information.

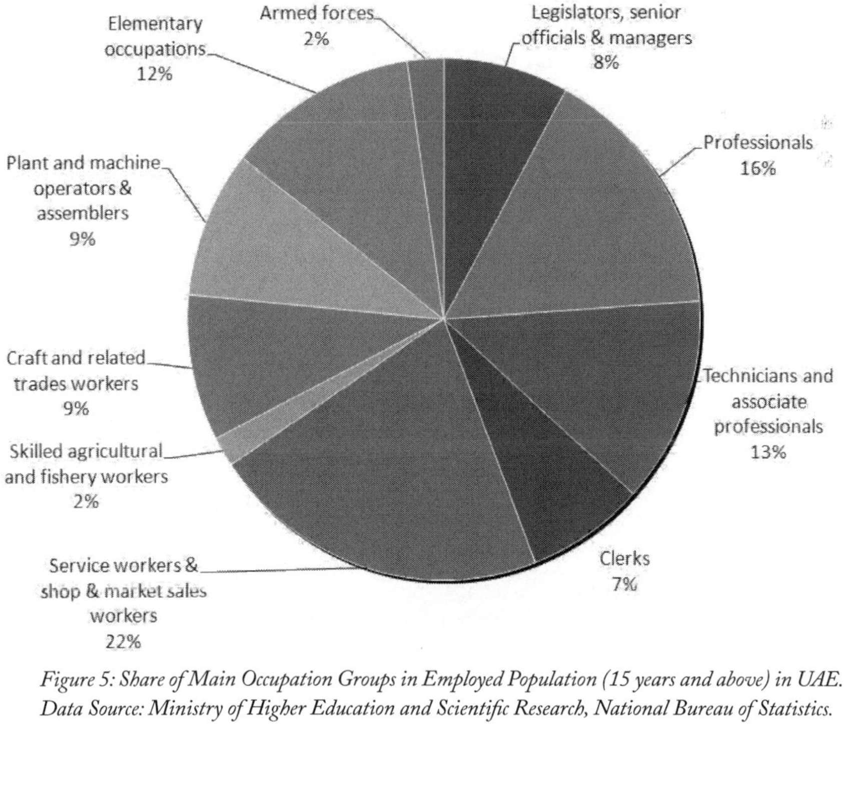

Figure 5: Share of Main Occupation Groups in Employed Population (15 years and above) in UAE. Data Source: Ministry of Higher Education and Scientific Research, National Bureau of Statistics.

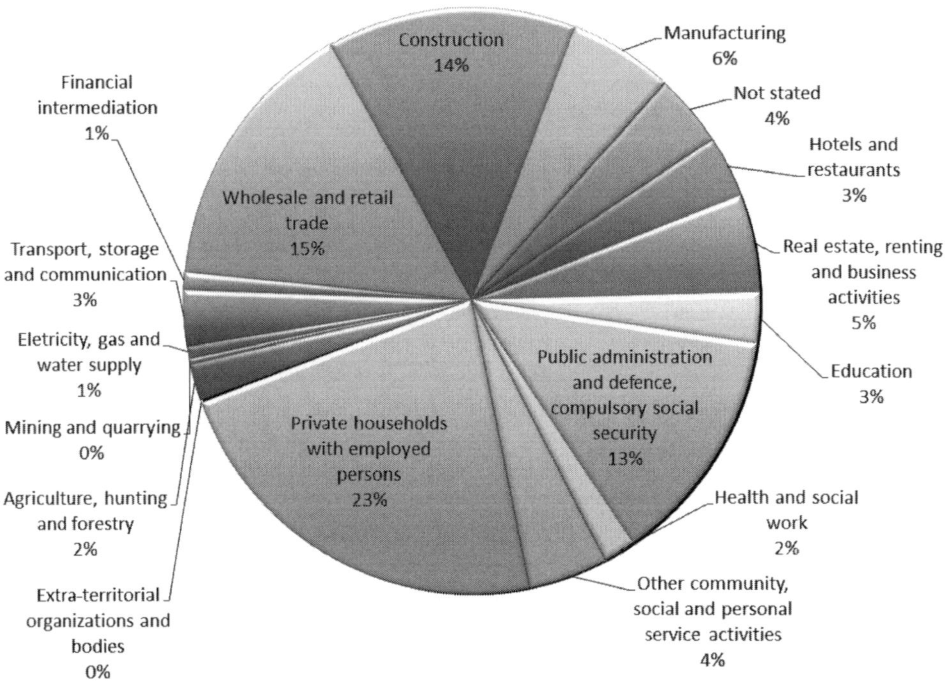

Figure 6: Share of Main Occupation Groups in Employed Population in Kuwait. Data Source: Public Authority for Civil Information, 2014.

Essentially, the vast majority of low skilled workers and resulting low labor productivity in the region is commonly associated with the easy availability of low skilled workers at internationally competitive rates and bound by flexible contracts wages. Although this has allowed the GCC to possess a highly elastic supply curve for foreign labour, allowing it competitiveness in the context of labour costs, it has been noted that this particularity has caused firms to typically substitute process and technology upgrades for low cost foreign labor (Hertog 2012, 2). Consequently, over the past few decades exceptionally large numbers of foreign workers have been employed for low skilled jobs causing overall GCC labor productivity to fall in the 1980s and stagnate since the 1990s (Hertog 2012, 2).

In recent times, labor productivity has declined further in many GCC countries (IMF 2013, 9). Over the period of 2000-2012, labor productivity, estimated as non-oil output per worker, decreased in Bahrain, Oman, Qatar, and the United Arab Emirates. Although labor productivity increased in Saudi Arabia and Kuwait approximately 1.5 per cent annually on average during the same period, in Kuwait it started to decline again since 2008 leaving Saudi Arabia as the only GCC states with rising labor productivity. This long-term decline in labor productivity in the overall GCC region contrasts sharply with comparable countries over the past decade, including Malaysia, Singapore, Hong Kong, Norway and Canada. It is likely that these declines are somewhat linked to the development strategy of large investments in infrastructure which have relied on these temporary and low-skilled migrant labor as well as possible inefficiencies in investment spending (IMF 2013, 20).

The Implications of Labor Market Characteristics in the Region: Key Challenges

These labor market characteristics hold various implications for the potential competitiveness of STI industries in the region. Firstly, larger volumes of low skilled labor over highly skilled labor and resulting low labor productivity indicates that at present GCC economies do not have the significant levels of human capital required to build strong and competitive STI industries in the region. This signals the imminence of a notable skills gap between the existing set of available talent in the region and the required talent in the future. Essentially, such a gap would appear to be characteristic of economies seeking to transition towards more knowledge based models and merely indicate the need for vigorous human capital development initiatives which, as previously discussed, are clearly a key priority for policymakers already. However, this gap specifically highlights the notably total low number of national workers (except for Saudi Arabia), in addition to low skills amongst nationals. This indicates that developing national human capital alone, will not be enough to supply the regions STI industries with the requisite amount and quality of talent due to the sheer lack of volume of national students and workers as seen previously (Figure 1). Consequently, human capital development in the region will have to involve largescale international recruitment of highly skilled talent, including students and workers, at least in the short to medium run. Furthermore, foreign talent such as experts, professors and professionals, will not only be critical for bringing in international knowledge and expertise to kick start the establishment of STI sectors in the short run but also for training and developing national students and workers in the long run. However, the international recruitment of highly skilled workers may be the most critical challenge in the region due to the temporary nature of employment contracts that may serve as a major deterrent to attracting and retaining highly skilled human capital in the GCC.

It is important to consider that as GCC expatriate populations differ significantly across nationalities, skill levels and income brackets, it is likely that the motivations that drive them to live and work in the GCC also differ along with their attitudes towards temporary work visas. Whereas workers from high income countries are likely to be drawn to the GCC due to the option of higher and tax free salaries, workers from lower income countries may also seek security and improved lifestyle conditions in addition to higher wages. This suggests that the temporary nature of GCC work visas may be more unattractive for skilled workers from lower income countries, who seek to achieve security and improved lifestyle conditions. However, as previously discussed, the largest expatriate populations in the GCC are from the neighboring lower income and/or politically volatile regions indicating that these visa structures are likely to be more unattractive to a significant majority of GCC expatriate populations.

As discussed previously, the nature of employment contracts in the GCC region is largely based on the sponsorship program which allows workers only 'temporary' or 'guest' status. Over the past few decades this system attracted large volumes of low skilled workers at competitive wages which have resulted in labor markets characterized by constant churn and low labor productivity. Thus far this system has been essentially suitable for GCC economies as they have been largely driven by sizeable, and in some cases dominating hydrocarbon sectors and other labor intensive sectors such as retail, construction and utilities. However, the foreseen transition of the GCC into more diversified economic models, which will be characterized by a greater emphasis on knowledge and information industries, will increase their need to attract and retain higher levels of skilled talent. For example, in the case of Abu Dhabi, sectors marked for further growth and expansion in the emirate include pharmaceuticals, biotechnology and life sciences, aviation, aerospace and defense and financial services amongst others. Undoubtedly these sectors will require more high skilled and specialized workers, in the fields of finance, engineering and medicine. Yet, employment mechanisms which are based on temporary contracts and must be renewed annually or every two or three years may render the GCC less attractive for higher skilled workers, relative to other economies by reducing residency and employment security. Additionally, the continuous churn of workers, caused by temporary employment contracts and residence permits, may also be unattractive for talent (Mahroum 2011, 50).

These challenges may notably disadvantage the region in attracting and retaining top talent particularly at a time when highly skilled workers are sought after globally and recent policies regulating immigration and entries of temporary workers in major OECD countries, which offer the additional advantage of reputable and long established STI industries, have been increasingly favoring highly educated workers (Mahroum 2011, 9).

Talent is observed to be more geographically attracted to areas which host other creative people and activities as opposed to areas where the critical mass of talent is low. According to agglomeration theory, industries with knowledge functions, such as science, technology and research and development, are found to create an 'innovation environment', which increases productivity and growth (Lambooy 1998, 5). Knowledge clusters, which attract large numbers of highly skilled workers, are found to attract more new workers due to 'agglomeration effects'. Furthermore, such activities are also found to thrive in more clustered environments thus creating a cycle of innovation. It is observed that talent inflows into such innovative environments rise incrementally and fortify and also strengthen knowledge assets which are already pre-existing in a certain area. According to Richard Florida, an American social and economic theorist, highly skilled human capital, technology and knowledge industries and openness, in the context of workforce deployment and immigration, form a 'creative labor class' and have strong positive effects on overall regional income (Florida 2001, 63). Creative industries therefore tend to compound the perceived attractiveness of a given economic area to firms and workers from knowledge and technology industries. It is worth noting here that high levels of skilled human capital and agglomerated knowledge communities often precede the arrival of local and foreign investment. For instance, many multinational enterprises have opened R&D centers in India, Ireland and Singapore due to the agglomeration of high levels of skilled workers in those locations (Mahroum 2011, 9). Furthermore, "the ability to attract and retain is as important if not greater than the ability to develop and produce domestic talent, particularly for smaller countries" (Mahroum 2011, 37). However, it is clear that the inflow of highly skilled workers is as critical to the productivity and competitiveness of an economy as is its existing stock of workers as inflows of foreign highly skilled workers will compete with domestic skilled workers thereby raising the overall quality of talent (Wolfe and Gertler 2004, 1084).

An example of such positive impacts of agglomeration can be seen in Canadian cities such as Toronto, Vancouver, Montreal and Calgary which are found to be leading centers in attracting and retaining highly educated and talented workers in a few specific STI industries (Wolfe and Gertler 2004, 1085). The significant presence of highly skilled workers in an area signals that these areas host the most innovative and interesting work in a given field and creates a reputation of innovation and creativity about them. Mechanisms to track these flows have also been formulated including the measurement of the inflow and in-migration of exceptionally highly skilled talent and well-known scientists. For example, recent analysis of such geographical attractiveness in Canada illustrates that cities such as Vancouver, Montreal and Toronto are exceptionally attractive for highly skilled talent and 'star scientists' particularly in the field of biosciences. Not only do these cities already host significant populations of skilled individuals in this particular field but also the most important advancements in the field. they therefore maintain a notable reputation for being the best places for highly skilled talent in this field. These trends are noticeable even on the firm level, firms that have high levels of renowned talented workers or experts experience significantly higher employment growth as other talented workers are attracted towards them (Wolfe and Gertler 2004, 1084).

In the GCC, recruiting top international talent in certain sectors has undoubtedly been challenging so far. For instance, attracting professors to teach at key educational institutions, which are increasingly being established across the region, has been found to be a notable challenge and has largely been associated with the low levels of existing experts in the region and the lack of a preexisting reputation for academia and research. For example, for Weill Cornell Medical College in Qatar, one of the first operational challenges was attracting and retaining top academic talent. Although it was acknowledged that recruiting talent is challenging in any location it was noted that it can become a more significant challenge "when recruiting to a nascent program in an environment that lacks many attributes normally found in an academically rich setting, including an established academic infrastructure and a critical mass of colleagues" (Hajjar and Gotto 2013, 73). Furthermore, according to a survey of Gulf medical schools, faculty recruitment and retention is considered one of the most critical challenges to the development of medical science education in the region.

It has been noted that high skilled immigration generates significant advantages for the host country by stimulating long distance 'social and economic linkages' with other science and technology centers thereby strengthening innovation and research capabilities domestically and building the regions 'economic dynamism'. According to AnnaLee Saxenian, Taiwanese and Indian engineers in Silicon Valley coordinated 'mutually beneficial ties between technology producers in Silicon Valley and 'state of the art' manufacturing and design expertise in international technology hubs such as the Hsinchu region and Bangalore and Hyderabad respectively allowing the region to maintain resilience during the economic crisis and flourish in the face of labor shortages in other areas of the country. Israel presents another example of transnational technology networks formed by returning engineers and venture capitalists. According to Saxenian 'the challenge of economic development will increasingly involve building such transnational social and professional linkages'. These examples highlight the critical relationships between immigration, trade and economic development and entrepreneurship. The economic impact of highly skilled immigrants extends from the direct effects, i.e. increasing the skilled labor supply, towards enhanced trade and investment flows. According to Saxenian, policy makers must acknowledge these relationships even though they are comparatively challenging to quantify (Saxenian 1999, 6).

Other Economies

It is important to note that many advanced and emerging economies have faced, and are currently facing, similar challenges in recruiting and retaining highly skilled talent due to the globally increasing demand for knowledge and information workers. In addition to developing domestic talent, as previously discussed, several economies have installed specific mechanisms over the past few decades to attract and retain STI workers such as the Green Card for IT specialists in Germany, the General Skilled Migration program in Australia and the Federal Skilled Worker Program in Canada. Each mechanism seeks to address the different requirements of each particular labor market and is, in many cases, continually amended to suit the changing needs of each market.

Germany

The German Green Card was established to address a significant skills shortage in Germany's IT sector in the late 1990's. During this time the domestic stock of human capital specializing in IT fields was so low that IT vacancies could take up to 20 weeks to fill and in some cases when vacancies could not be filled projects would have to be declined. This shortage of skills was found to be weakening the innovation strength and competitiveness of German industry and causing firms to consider relocating. In order to address this shortfall, in 2001 a 'Green Card' initiative was launched by Chancellor Schroder in order to make it possible for specialists from non-EU countries to be employed in an IT activity for 5 years. Foreign workers who were eligible included specialists in information and communication technology i.e. specialists in software development, multimedia development and programming, circuits and IT systems development and IT consulting amongst others. The Green Card initiative was particularly relevant to foreigners from outside the EU and concerned people with a degree from a university of polytechnic in information and communication technology but was also applicable to foreigners graduating from German universities and polytechnics, who could previously not stay in Germany after graduation.

The results of the Green Card initiative indicated that firms rapidly saw positive effects. The initiative allowed firms to fill vacancies within 6 weeks, less than half the time it took before the introduction of the initiative, many IT firms increased their investments in Germany and were able to expand operations and plans to move business activities to other countries were withdrawn. Additionally, an increase in overall employment of green card employees and local workers, was seen. In 2000, when the initiative was launched, approximately 17500 jobs were created in 6 months, out of which 5000 were for green card employees and 12500 were for local workers (Werner 2002, 21) . Due to the positive effects on competitiveness and innovation of the Green Card initiative, a new Immigration Act came into force in 2005 which made it easier for highly qualified people to be granted work permits and permanent residence from the onset as opposed to the 5 year permits granted under the Green Card initiative. Professions that were most sought after include biologists, physicists, chemists, engineers, professors and scientific personnel in high technology areas. Moreover, family members of highly skilled workers who obtain a visa were also given the right to work in Germany (Work Permit 2014). More recently, the EU 'Blue Card' has been installed in Germany that allows highly skilled workers to move within the EU. Although these mechanisms have met with mixed response in terms of overall political attitudes towards immigrant integration, it is important to note that the inflow of highly skilled workers into Germany from within the EU and from other regions has been an important advantage for the country's competitiveness in STI sectors (Kinkartz 2012).

Australia

Australia has also experienced notable skill shortages in various industries over the past few decades. Historically, skills shortages in fields such as mining, engineering and information technology were identified as the most critical to industry growth. However, in recent times skills shortages in other growing sectors, such as health diagnostics, have also gained significance (Department of Employment 2014). The federal government has undertaken various measures in response over the years.

The Australian government created the General Skilled Migration program (GSM) to draw and retain skilled and experienced migrants in sectors with imminent skills shortages. Through the GSM, skilled migrants can apply for migration from within Australia, from outside Australia and if they are graduating from an Australian university. Furthermore, there are also specific mechanisms installed to attract successful entrepreneurs and businessmen to reside in Australia and create new business or develop existing ones. It is important to note that the Skilled Migration program creates options for skilled workers who want to live in Australia and who do not have an employer sponsoring them. These include options for skilled people applying as independent migrants as well as those sponsored by a relative, or nominated by a State or Territory government. In 1996, the 457 Skilled Migrant Visa was introduced that allowed employers to hire overseas workers to fill nominated skilled positions which they could not fill domestically. Temporary skilled migrants can be on 457 visas for up to four years, with an option to move to Permanent Residency after that period. In 2011, the number of primary 457 visa holders in Australia was 72,030, an increase of 5.3 per cent in comparison with the same date in the previous year (Bahn and Cameron 2013, 28). In 2011, limits on sourcing foreign workers were raised and Western Australia was rezoned as a regional center thus allowing an unlimited supply of skilled overseas labor for the state.

Canada

For more than three decades, Canada 'pioneered' the use of a comprehensive skill-based points system for selecting economic migrants who are highly educated and possess skills deemed applicable to the domestic labor market (Ferrer et al. 2012, 1). In 2002, the current Federal Skilled Worker (FSW) program was introduced, in order to more effectively select immigrants who are likely to succeed economically in the longer run. The program amends the earlier immigration model which sought to cater specifically to skill shortages in certain sectors and industries by allotting points for those occupations by basing selection purely on human capital and not on occupation. The model was designed to "maximize the long-term potential of economic immigrants in an increasingly complex labor market and knowledge-based economy" (Ferrer 2012, 11) and seeks to improve the economic outcomes of skilled worker immigrants and maintain the quantity of skilled worker immigrants. Skilled workers under this system are selected according to their level of education, language ability, work experience, age, arranged employment and adaptability amongst other factors.

In more recent times, Canada has expanded its mechanisms to recruit internationally by introducing the Temporary Foreign Workers Program. While most temporary foreign workers have been hired to address a specific, short-term labor shortages, some temporary foreign workers who initially came to fill a temporary vacancy have transitioned to permanent residence if they meet requisite criteria. For example, the Canadian Experience Class is open to foreign nationals who have been working full-time in Canada as trades people or in managerial or professional occupations and meet certain other requirements. Other foreign workers may qualify through the Provincial Nominee Program for permanent residence in Canada. These mechanisms allow workers whose skills continue to be in demand and are already familiar to Canada can contribute to the economy in the long run (Government of Canada 2012).

Important Lessons for GCC Labor Markets

The models from other countries provide various lessons for GCC labor markets.

Sourcing Foreign Talent is an Important Policy Issue in the Development of STI Industries

Industrial economies recognize the importance of the share of foreign workers in the supply of talent to their STI industries. Accordingly, they have a very wide range of mechanisms to facilitate domestic companies in sourcing foreign talent. In the context of the GCC, it must be acknowledged that implementing such policies and mechanisms to address labor market demands will be essential for fostering STI sectors. Efforts to develop talent nationally will have to be in parallel with talent inflows, particularly in the short run as skilled workers will be required to train domestic students and staff.

It is commonly argued that the strength and competitiveness of a given economy is shaped chiefly by its institutional strength and that workers are merely guided by institutional frameworks and norms. Institutions, in this regard, refer to a country's "norms, rules of conduct, and generally accepted ways of doing things" and "economic institutions are formal and informal rules of the economic game" (Todaro and Smith 2006). Nevertheless, it is important to note that although institutions do constitute a critical aspect of an economy's governing framework they cannot substitute for the knowledge and expertise embodied in people. Although institutions do eventually form the accepted guidelines of economic 'behavior' for a given economy and therefore workers would likely adapt. However, talented workers hold individual expertise which is built on incrementally over time and would theoretically be disrupted if it was constantly replaced.

Mechanisms are not Static

Mechanisms to source talent notably vary across economies and over time. They are essentially tailor made to suit the specific labor demands of the particular economy or sector and are continually altered to adapt to changing labor market demands and according to visible results. For instance, Germany installed a visa mechanism that was initially targeted at addressing a specific skills shortage for a set period of time but was eventually expanded to offer longer stay options for foreigners due to its effectiveness in strengthening domestic industries. On the other hand, Canada and Australia, who have traditionally offered longer-stay and permanent residence options and have taken into consideration the potential suitability and adaptability of the given worker to adjust the country and society, have been increasingly installing more temporary forms of international recruitment. This indicates that policymakers in the GCC can design mechanisms to suit specific labor market demands and alter these as results are mapped.

Temporary Contracts are Effective Mechanisms for Certain Labor Markets

Although it may be indicated that temporary employment contracts themselves do not necessarily deter skilled employees, it is likely that this is only the case when such contracts have been coupled with the right incentives. In advanced economies such contracts have been coupled with the draw of well-established STI industries and options for longer stay at the end of residency terms. Furthermore, temporary contracts in such economies have largely been offered at longer stay periods than those offered in the GCC. These factors indicate that although temporary contracts themselves may not necessarily deter highly skilled workers their structures must be adapted to changing labor market demands and reduce residency and employment insecurity. This will also be crucial for fostering an environment that is conducive to research and innovation as research and innovation are founded upon long-term initiatives and projects which require a gradual accumulation of information and knowledge. A workforce that is susceptible to constant turnover may be less suitable for research and innovation.

Conclusion

In conclusion, it can be seen that it will be critical for policymakers in the GCC to recognize the negative incentives created by existing labor market structures and their implications for nascent STI industries in the region. It will also be important to reconsider the importance of sourcing foreign talent as an integral element of building national STI strategies. Presently it appears that, the imminence of a skills gap between the low skilled-low productivity workforce, which characterizes labor markets in the region in the present day, and the high skilled-high productivity workforce required for STI industries in the future, is not perceived as very problematic. It may be assumed that as the levels of human capital in various areas start accumulating in the region, with the ongoing development of STI education and training facilities, the GCC will also eventually become a notably attractive region for highly skilled talent. Businesses across all sectors in the region have already observed that creating a skilled workforce should be a top priority in the region as the current availability of skills is inhibiting fast growth (PWC 2014, 16). In the context of the STI sector this challenge will be particularly great due to the lack of preexisting STI industries and assets in the region and also due to simultaneously increasing demands for skilled workers from emerging and advanced economies.

Fundamentally, in order to start addressing the challenges involved in building an STI workforce it will be critical for GCC policymakers to map and study talent inflows and outflows in region and enhance their understanding of regional labor mobility. Research focusing on the international mobility of highly skilled workers is being widely undertaken by several advanced and emerging economies in order to understand the supply side of talent. For example, a Skills Research Initiative was established in Canada in 2003 between Industry Canada, Human Resources Development Canada and the Social Sciences and Humanities Research Council. This initiative sought to fuel 'policy-relevant' research on skills in the context of international mobility of highly skilled workers, employer supported training and other critical aspects of labor market supply. In Australia, the Department of Employment carries out annual statistical research on labor market demands and highlights skills requirements and shortages in order to facilitate incoming worker selection. Research on GCC talent flows is exceptionally sparse and must be stepped up in order to develop methods for more accurate assessment of global trends in the migration of highly skilled labor and determine the GCC's position in this global market relative to other industrialized countries. This will allow the GCC to improve its understanding of the fundamental factors underlying the increased migration of highly skilled labor, especially among developed countries, and improve its understanding of the economic costs and benefits associated with the international mobility of highly skilled labor and the main factors influencing these costs and benefits. In the long run, it will be important to separate the perceptions of mechanisms for immigration and permanent residence from those of supplying highly skilled labor to national STI industries through more stable and attractive channels. The confusion between the previous two structures can potentially hinder the competitiveness of the ongoing development of STI sectors in the region.

Lastly, it is indicated that such national strategies (that would likely vary across GCC countries) must involve the creation of mechanisms to facilitate international recruitment in a manner that renders them more favorable in terms of residency and employment security. In recent times, emerging technology hubs such as Taiwan have started to install similar policy mechanisms to strengthen incentives for non-national talent. Installed measures will allow prospective foreign investors who have raised NT$2 million (USD 64,000), or more from domestic or foreign startup platforms or have secured related patents, to be granted residential visas which can be extendable for an additional two years after the investment project begins. Presently, according to Taiwanese existing immigration laws foreign nationals have to invest at least NT$6 million (USD 190,000) in order to obtain residency in Taiwan. Furthermore, medium-level technicians, in designated fields will be issued renewable three-year residency visas, similar to treatment for white-collar workers in Taiwan and to existing immigration structures in Singapore (Taipei Times 2015). The installation of such structures in emerging technology hubs not only represents the changing attitudes towards foreign workers and the significance placed on sourcing foreign talent but also the varying and targeted nature of mechanisms which seek to address specific policy goals and smooth out labor market inefficiencies. A consideration of similar mechanisms will be essential for the GCC in order to be able to keep up with advanced economies that have already recognized the importance of foreign talent and continually attract large numbers of highly skilled workers. Temporary employment programs installed in Canada and Australia indicate that temporary contracts themselves are not necessarily deterrents for highly skilled workers. However, they must be coupled with the right incentives and must allow some sense of residency and employment security regardless of their temporary nature. In Canada and Australia, such contracts have been coupled with the draw of existing knowledge clusters and therefore hold different connotations than those for economies with nascent STI sectors. Moreover, they appear to be of notably longer periods than those in the GCC region and workers are offered longer stay options at the end of their terms. Although high levels of remuneration are found to be attractive to foreign workers they are not found to be notably attractive to senior level and high talent workers, particularly within high-tech sectors, which are largely found to be attracted to more professional incentives (Government of Canada 2012).

Essentially it will be important to understand that longer-term temporary employment contracts are not synonymous with more complex issues such as immigration or citizenship. They are targeted and effective policy responses to national economic requirements. Consequently, policymakers in the GCC must also consider forming specific mechanisms to source international talent at periods longer than the existing nature of annual residence and visa contracts in order to increase residency and employment security and start building a larger supply of domestic capabilities.

References

Abouchakra, Rabih et al. 2008. *"Economic Diversification The Road to Sustainable Development"*. Booz & Co.

Abu Dhabi STI Vision. 2014. Abu Dhabi Technology Development Committee.

Alsodais, Sami. 2013. *"Science, Technology & Innovation in Saudi Arabia"* WIPO Magazine

Babar, Zahra. 2013. *"The State and Innovation in the Gulf"*. Centre for International and Regional Studies, Georgetown University, Qatar.

Bahn, Sussane and Cameron, Roslyn. 2013. *"Sourcing Specialized Skilled labor in the Global Arena: A Change in the Way we View Work in Australia?"* Australian Bulletin of Labor.

Brinkley , Ian et al. 2012. *Kuwait and the Knowledge Economy*. Kuwait Foundation for the Advancement of Sciences, London School of Economics, 2012.

Brown, Phillip, Lauder, Hugh and Ashton, David. 2008. *Education, globalization and knowledge economy*. Teaching and Learning Research Council, UK.

Department of Employment. Australian Government. 2013 *"Statistical summary of skill shortage research in 2013"*

Educate to Innovate. 2014. "Education for K-12 Students" The White House. Last Modified May 16 http://www.whitehouse.gov/issues/education/k-12/educate-innovate

Ferrer, Ana, Picot, Gernett and Riddell, Craig. 2012. *"New Directions in Immigration Policy: Canada's Evolving Approach to Immigration Selection"*. Canadian Labor Market and Skills Researcher Network

Florida, Richard. 2001. *"Bohemia and Economic Geography"*. Journal of Economic Geography, Oxford.

GCC: Kafala, UAE. Migration News. 2012. University of California, David. Accessed May 16 2014, http://migration.ucdavis.edu/mn/more.php?id=3740_0_3_0]

Government of Canada, Industry Canada, 2012. "International Mobility of Highly Skilled Workers". Accessed May 16 2014 https://www.ic.gc.ca/eic/site/eas-aes.nsf/eng/ra02070.html

Government of Canada. 2012. *"Temporary Foreign Worker Program, Citizenship and Immigration in Canada"*.

Hajjar, David and Gotto, Antonio M. *"Launching of an American Medical College in the Middle East: Educational Challenges in a Multicultural Environment"*. International Journal of Higher Education.

Heinz, Werner. 2002. *"The current "green card" initiative for foreign IT specialists in Germany"* International Mobility of the Highly Skilled, OECD.

Hertog, Steffen. 2012). *National employment, migration and education in the GCC*. Berlin: Gerlach Press.

International Monetary Fund. 2013. *"Labor Market Reforms to Boost Employment and Productivity in the GCC"*. Annual Meeting of Ministers of Finance and Central Bank Governors.

Isalkar, Umesh. 2014. *Intel India launched National STEM acceleration program.* The Times of India
Katsomitros, Alex. 2013. "The global race for STEM skills". Accessed May 16 http://www.obhe.ac.uk/newsletters/borderless_report_january_2013/global_race_for_stem_skills
Kinkartz, Sabine. "*German 'Blue Card' to simplify immigration*". Deustche Welle. Accessed May 18 2014.
Lambooy, Jan. 1998. "*Economies of agglomeration and spatial development*" University of Utrecht
Lori, Noora. 2012. "*Temporary Workers or Permanent Migrants?*" Center for Migration and Citizenship.
Mahroum, Sami. 2011. "*Develop, Deploy and Draw: Backing Abu Dhabi's 2030 Total Human Capital Potential*" INSEAD.
Pradhan, Samir. 2010. "Foreign Labor in the Gulf. On the Move: Migration Challenges in the Indian Ocean". The Henry L. Stimson Center.
PwC Middle East. 2014. *"Fit for the future Building on today's optimism to prepare for tomorrow, CEO Survey".*
Saxenian, AnnaLee. 1999. *"Silicon Valley's New Immigrant Entrepreneurs".* Public Policy Institute of California
Strategy, PwC. 2012 *"Economic Diversification: The Road to Sustainable Development"* Accessed May 16 2014 http://www.strategyand.pwc.com/me/home/thought_leadership_strategy/40007409/40007869/43406964
Taipei Times. 2015. "*Attracting the right global talent.*" Accessed December 2016. http://www.taipeitimes.com/News/editorials/archives/2015/03/09/2003613100
Todaro, Michael and Smith. 2006. Stephen. *Economic Development 9th Ed.* Pearson Education Limited.
Wolfe, David and Gertler, Meric. 2004. *"Clusters from the Inside and Out: Local Dynamics and Global Linkages".* Urban Studies
Work Permit. Introduction to Immigration to Germany. Accessed 10 May 2014. http://www.workpermit.com/germany/employer1.htm

Note

1 Indicators represent: UAE: Distribution of Population (10 Years +) by Educational Status according to Emirate, Saudi Arabia: Labor Force (15 Years and Above) By Education Status and Sex, Qatar: Population (15 Years and Over) By Age Groups and Educational Status, Kuwait: Education Status, Bahrain: Population (15 Years +) by Highest Educational Level and Nationality. Sources: Bahrain Central Informatics Organization (CIO), Census 2010, Kuwait Public Authority for Civil Information)PACI), Dec. 2012., Qatar Statistics Authority (QSA), Quarterly Labor Force Survey 2013 (Q4), Saudi Arabia Central Department for Statistics and Information (CDSI), Labor Force Survey 2013, Round 2, UAE National Bureau of Statistics (NBS), Labor Force Survey

4

Key Factors in the Tertiary Educational Trajectories of Women in Engineering: Trends and Opportunities in Saudi Arabia, the GCC, and Comparative National Settings

Jennifer DeBoer and Ashley Ater Kranov

Introduction

The participation of women in engineering undergraduate programs and in the workforce in numerous Gulf Cooperation Council (GCC) countries stands in contrast to the systematically low proportions of female engineers in the United States (US) and numerous western European countries. Across most of the GCC nation-states, women attain degrees at the tertiary level at near parity to male students, and they participate in engineering undergraduate studies at significant rates (Charles 2011). Although this varies by engineering discipline and by GCC country, and although this balance is less equal in some sectors of the workforce, the participation of women in the technical innovation enterprise has been an important component of growth in the GCC (Al Masah Capital Management Limited 2012; Shehadi et al. 2011).

This has not always been the case. Women's participation in economic, educational, and political arenas in GCC countries has been historically low, primarily due to socio-cultural factors influencing laws and policies that have hindered involvement in the public sphere. Over the last quarter century, as attitudes and policies have changed, so has women's representation in these arenas. Yet, in spite of these changes, the female participation rate in the general workforce in the GCC stands at 27%. Again, their overall participation remains at just over one-quarter of the workforce even though women graduate from college or university at higher rates than men in the GCC and also at higher rates than women outside the GCC (Shehadi et al. 2011). Women in Kuwait, Qatar, the UAE, Oman, and Saudi Arabia represented 67%, 62%, 60%, 59%, and 57%, respectively, of their nation's university graduates in 2009 (UIS 2015). As a comparison, in the same year, women comprised 58% of graduates in the US and 57% in the United Kingdom (UIS 2015). This pool of highly educated women in the GCC, including many engineers, is an underutilized resource that, if leveraged, could greatly benefit the region's economic, political, and societal spheres.

Why is the study of women's participation in engineering across countries important? Further exploration into the multi-faceted core causes of gender disparity in education and the workforce are necessary because such disparities:

1. limit the educational, career, and life opportunities of both boys and girls;
2. promote "separate but equal" distribution principles that frequently do not result in equal pay or power; and
3. prevent women from filling the growing global shortage of technical expertise (Charles and Bradley 2009).

GCC countries are in the unique position of having relatively new educational institutions in place and, therefore, may be well situated to build educational policies and structures that respond to the needs of their citizens, the region, and the global economy. Women's participation in the GCC's growth trajectory in both the economic and educational spheres is significant (Shehadi et al. 2011). Capturing a current, cross-sectional measure of women's engagement as engineers is one step in leveraging existing expertise in the GCC and MENA regions, cultivating existing success pathways, and developing new ones (Al Masah Capital Management Limited 2012).

In the next section of this chapter, we delve into trend information and current data in the GCC countries to illustrate women's educational participation during the tertiary schooling and in engineering and STEM. This includes data on educational performance and attitudes, incorporating relevant pre-university information, as well as labor force statistics. Section three of this chapter provides a succinct review of the literature on gender parity in STEM fields worldwide. In section four, we then provide "country brief" case studies of one GCC country (Saudi Arabia) and a selection of non-GCC comparison countries, including the US. Section five discusses the implications of new educational structures for women's educational and occupational opportunities in the region. We conclude with a brief discussion of implications, policy recommendations, and remaining opportunities.

Relevant Trend Information and Current Data on GCC Countries

We begin with available data that illustrate trends in women's participation in engineering studies in the GCC countries. In addition, we look at women's enrollment and attainment in the tertiary (and, for added context, pre-tertiary) education system overall. We then look more closely at women and girls at early stages of their engineering educational trajectories in terms of their STEM academic performance, their perception of the field, and their self-efficacy and confidence in the field. Next, we discuss performance during tertiary education where engineering domain learning is focused, as well as student perceptions of their motivations towards learning engineering content. Finally, we discuss available information on the labor force in the GCC in terms of STEM researchers and gender ratios in STEM field jobs.

For all of these metrics, we draw on existing international and comparative datasets. The Organisation for Economic Co-operation and Development (OECD) runs the Programme for International Student Assessment (PISA), which assesses math, science, and reading literacy for 15-year-old students (ostensibly in their last year of mandatory formal education). To complement the PISA data, we present descriptive information from United Nations Educational, Scientific, and Cultural Organization (UNESCO)'s Institute for Statistics (UIS), including women's participation in STEM research. Finally, we present corresponding data from the International Labor Organization (ILO), where available. These data not only contextualize our descriptions of participation in engineering for women in GCC countries, but they also inform the choice of countries to study in our international, cross-comparative investigation of women's engineering study and practice.

The three sections of Table 1 (a-c) provide an overview of key metrics that we have compiled from the UNESCO Institute for Statistics digest and UNESCO World Atlas of Gender Equity. Table 1.a. highlights enrollment and dropout metrics for secondary and tertiary education for both females and males (for comparison). Table 1.b. gives attainment figures for the current adult population in GCC countries, where data are available. Finally, Table 1.c. provides enrollment, attainment, and research and development (R&D) labor force participation figures specific to engineering and STEM fields in GCC countries. We follow the International Standard Classification of Education (ISCED) 2011 guidelines to categorize educational levels and to define secondary (ISCED 2, 3) and tertiary/postsecondary schooling (ISCED 6 and up; UNESCO 2015).

Table 1.a. Enrollment in General Education (Secondary/Tertiary)

Enrollment in General Education (Secondary/Tertiary)						
Country / Percentage (%)	Dropout Rate (lower secondary)		% Students in Secondary who are Female			Female Tertiary Enrollment (% of student body)
	F	M	Lower	Upper	All	All Fields
Bahrain	1.9	2.8	48.3	50.0	49.1	59.2
Kuwait	0.9	15.5	49.1	52.3	50.3	63.9
Oman	1.45	1.7	48.9	49.0	48.8	52.8
Qatar	3.7	10.9	49.3	48.7	49.0	65.1
Saudi Arabia	16.1	5.4	47.1	46.1	46.6	49.3
United Arab Emirates	4.1	6.7	48.7	49.5	49.0	55.4

Source(s): UIS Statistics, 2015.

Table 1.b. Attainment of ISCED Qualification Levels

Attainment of ISCED Qualification Levels (Population 25 and older, by ISCED 2011)						
	6 (Bachelor's or equivalent)			8 (Doctoral or equivalent)		
Country / Percentage (%)	F	M	MF	F	M	MF
Bahrain	10.0	7.7	8.5	0.1	0.2	0.2
Kuwait	11.1	6.8	8.3	0.3	0.5	0.4
Oman	11.3	15.3	14.0	-	-	-
Qatar	34.5	17.1	20.5	-	-	-
Saudi Arabia	0.6	1.3	1.0	-	-	-
United Arab Emirates	25.1	15.7	18.0	-	-	-

Source(s): UIS Statistics, 2015; ISCED 2011.

Table 1.c. Enrollment, Attainment, and Labor Force Participation in Engineering/STEM

Enrollment, Attainment, and Labor Force Participation in Engineering/STEM					
	Engineering, Manufacturing, and Construction		**Engineering & Technology Researchers**	**Natural Sciences Researchers**	**Total R&D workforce**
Country / Percentage (%)	**Female Tertiary Attainment (% of whole graduating body)**	**Female Tertiary Enrollment (% of student body)**	**Females as % of total researchers**	**Females as % of total researchers**	**% Female**
Bahrain	25.7	21.1	-	-	
Kuwait	25.0	43.4	28.4	28.3	40.7
Oman	52.7	37.1	12.1	31.5	22.4
Qatar	27.4	34.3	-	-	-
Saudi Arabia	3.4	5.9	-	-	-
United Arab Emirates	31.1	35.1	-	-	-

Source(s): UIS Statistics, 2015.

As we compare overall enrollment and attainment metrics for the GCC countries, we can see notable diversity across the six countries in their outcomes for women. In Qatar, about twice the proportion of women as men have attained the ISCED equivalent of a bachelor's degree, and females have a lower dropout rate than their male counterparts. By contrast, Saudi Arabia has a higher female dropout rate and a slightly lower female proportion of secondary enrollment (about 46% in upper secondary). Although there is a great deal of variation in the proportion of the students enrolled in engineering programs who are women, many of the GCC countries show female enrollment in these programs at rates higher than in North America and Western Europe (e.g., women as 43.4% of the student body in engineering programs in Kuwait). Kuwait and Oman also provide data on engineering and STEM researchers. In both contexts, women make up a notable group of the researchers, whether in engineering and technology, natural sciences, or the broader R&D workforce.

Participation: Enrollment and Attainment

First, we present information on women's participation in secondary and tertiary education overall (all disciplines, including engineering). Statistics in the following two paragraphs are drawn from the UIS Data Centre (Data Centre 2015) and its records across countries, years, and topics, except where specified. Across many of the GCC countries, drastic increases in overall enrollment, related to the push for Education for All (UNESCO 2014), have bolstered the proportions of the school-going population of females. For example, female enrollment in Bahrain has trended upward with the dramatic overall increase in the gross enrollment ratio—up to 36 times previous levels (UNESCO 2012). For the adult population of at least 25 years of age in all GCC countries, a higher proportion of women than men have attained the highest level of tertiary education, International Standard Classification of Education (ISCED) levels 6 and up (postsecondary, tertiary) (ISCED 2015). In some cases, this difference is nearly double; for example, in Qatar, 33.9% of women in that age range have attained ISCED 6+ qualifications as opposed to 17.9% of men. Other empirical work corroborates this tendency in Qatar. Felder and Vuollo (2008) describe the beneficial prospect of the inclusion of Qatari women in the workforce. They describe women in Qatar as better educated than Qatari men, with great ambition for advancement and success in their academic and professional careers. Qatari women are increasingly joining the workforce, and, according to Felder and Vuollo, display tendencies for higher perseverance in their employment alongside windows of motherhood. Furthermore, the authors note that Qatar has enhanced opportunities for women through important changes in political, educational, and social areas. However, female economic participation is still on the low side in Qatar, even with recent increases and numerous improvements to the policy environment.

Despite drastic increases in enrollment across the GCC, there are numerous instances like Qatar where the profile of women's enrollment and attainment is still mixed. For the latest year of attainment data that is available for the adult population in each respective GCC country, the percentage of women with the highest ISCED level of qualification is only less than men in Oman, and there, only slightly so (11.3% women vs. 15.3% men). Still, on the other hand, in Oman, there are slightly more women with ISCED 4/5 (postsecondary, non-tertiary) qualifications. And, there are slightly higher proportions of women who are currently enrolled in ISCED tertiary levels 6+, which may reflect current changing trends in that country. Indeed, the rising profile of pre-tertiary enrollment and attainment may explain women's participation in tertiary education and indicate where changes are beginning to occur. Some of the remaining gender gaps in adult attainment may be attributable to persistent differences in retention and completion in lower education levels. For example, trends in Oman and in Qatar show higher numbers of female repeaters than males in secondary education (UNESCO 2012). Differential dropout rates plague female students in Saudi Arabia

(11% dropout rate for females vs. 2% for males), and, although Saudi's secondary gross enrollment rate (GER) is high for both females and males, it continues to be notably higher for males (103% vs. 90%). As a counterpoint, male dropout is much higher in Qatar (10% male vs. 3% female). Qatar has notably higher secondary GER for females than for males (107% female vs. 73% male, one of the largest female overrepresentations around the world), which then translates into higher tertiary numbers (UNESCO 2012).

Upon closer inspection of enrollment and attainment by discipline, we see that there is strong and growing representation of women in engineering and STEM study and work. In fact, higher proportions of the female tertiary student cohorts are enrolled in science in Saudi Arabia (16.3% of female vs. 12.7% of male students), in the UAE (8.2% of female vs. 7.9% of male students), and in Oman (24.7% of female vs. 18.3% of male students; UIS 2015). This may be related to the overall greater participation in tertiary education for female than male students in Oman, UAE, Qatar, and Saudi Arabia. There are still higher proportions of male students in GCC countries in engineering, manufacturing, and construction, however, it is not possible to disentangle these three complex and different fields from each other because the existing data do not differentiate between the disciplines (UIS 2015). And, although they are not the majority, females make up a significant portion of the STEM research labor force, whether in engineering and technology (Kuwait 28.4%, Oman, 8.9%); natural sciences (Kuwait 28.3%, Oman 24.8%), or total R&D (Kuwait 40.7%, Oman 25.7% [in 2002]). Throughout the engineering and STEM pipeline in the GCC, female students and practitioners participate at strong and growing rates.

Performance and Attitudes Towards STEM

Here, we summarize performance in and perceptions of STEM for nationally representative samples of 15 year olds who are poised to transition into higher education or the workforce. The PISA system also measures students' intrinsic and instrumental motivation to study mathematics. In the UAE and Qatar, boys report much higher intrinsic motivation to study math. By age 15, for many female students in GCC countries, there is an observable drop in positive attitudes towards math and science from previous high scores on earlier surveys. For example, in UAE and Qatar, there is a small but observable gender gap that favors boys in relevant attitudes towards learning such as "openness to problem solving", which PISA defines as an attitude toward general problem solving, even accounting for differences in math performance (PISA 2012). And, in the UAE, there is an increased likelihood of boys scoring among top performers in problem solving as opposed to girls (14% increase in relative likelihood). However, UAE boys are also more likely to score among the lowest performers (18% increase), which mirrors what has been observed in other national contexts for math/science performance of boys vs. girls (that is, higher variance in performance among boys, with similar mean achievement; Hyde et al. 2008). In fact, UAE girls on average outscore boys by 26 points on problem solving, by 28 points on science, and 5 points on math (PISA 2012). Similarly, in Qatar, girls on average outscore boys by 35 points on science and 16 points on math (PISA 2012). And, controlling for math performance, girls across the GCC still report higher instrumental motivation to study math (PISA 2012).

Review of the Literature on Gender Parity in STEM Fields Around the World

How do the trends we describe for women in engineering in the GCC fit into the worldwide picture? The state of women's engineering participation in the US and in the other contexts cited in this chapter, including a number of Gulf Countries and other majority-Muslim countries, can be situated in the broader literature on women's participation in North America, Europe, and a handful of other contexts. Numerous research teams have tracked the global expansion of women's enrollment and subsequent participation in STEM fields over time. Women's enrollment in higher education appears to be growing, to greater or lesser extents, in many places. For example, Ramirez and Wotipka (2001) conducted a cross-national study that demonstrated a significant increase in the enrollment of females in the fields of science and engineering in higher education. This was apparent globally across a span of twenty years, from 1972 to 1992. They used UNESCO Statistical Yearbooks to investigate the magnitude of and change in male and female enrollments in engineering across all regions of the world—as absolute levels, as percentages of respective totals in higher education, and most importantly, as women's share of higher education enrollments in science and engineering. On average, overall, women's enrollment globally in science, technology, and engineering fields more than tripled over that 20-year period (for the sampled cases in their study). Furthermore, women's share of the enrollment in science and engineering fields increased in all countries (mean change: from14.5% to 20.7%). This change is not consistent across countries, which we discuss in more detail below. In Europe, for example, Eastern European women's participation only slightly increased, from 24.8% to 25.6%, while women's participation in Western Europe increased from 17.2% to 23.8% between 1972 and 1992. In fact, scholars investigating women's participation in the UK and Europe more broadly note that growth in tertiary engineering enrollment has not actually changed the persistent barrier in terms of retention in the field or advancement to higher-level positions in STEM organizations (academic, industry, or otherwise) (Bebbington 2002).

Considerable global scholarly and policy attention has been directed towards studying the factors that explain this underrepresentation of girls and women within STEM fields, in particular in engineering, and largely in North America and Europe (Hill, Corbett, and St. Rose 2010; Cech et al. 2011; Ceci and Williams 2009; Charles and Bradley 2002; Charles and Bradley 2009; Lagesen 2008; Luke 2002; Riegle-Crumb et al. 2012; Xie and Shauman 2005). Several factors have been identified as shaping the uneven distribution of women and men across engineering fields of study, including: women's overall status within societies, the structure of national educational systems, and international and national cultural beliefs and norms regarding women. These factors operate through a number of mechanisms to produce gender-differentiated participation. These mechanisms might be processes that affect individual decision-making such as stereotype threat, gender-based differences in cultural and social capital, and the notion of engineering as a masculine-gendered field. Even macro-level forces such as societal and global cultural norms and the logic and structure of educational systems themselves play a role (Charles and Bradley 2009; Faulkner 2000; Marchand and Taasoobshirazi 2013; Morgan 1992; Seymour and Hewitt 1997).

Sociologists Charles and Bradley are among the only researchers to have conducted a robust study of gender segregation in a variety of fields across 44 countries, using their Gender Essentialist and Self-Expressive Value Systems Framework as a method for data analysis and interpretation (Charles and Bradley 2009). The study's counter-intuitive findings suggested that gender segregation in fields traditionally regarded as "masculine," such as engineering, is much more pronounced in high GDPPC (Gross Domestic Product Per Capita) countries like the US. They found that cultural beliefs in fundamental and innate gender differences express themselves in curricular and career choice more prevalently in economically developed countries where self-expression and individualism in curricular and career choice is highly valued, in contrast to less economically developed countries, where curricular and career choice is more frequently influenced by economic and prestige factors.

In another empirical example, as part of the pan-European project "WomEng", Sagebiel and Dahmen (2006) find both explicit and "latent" discriminatory or masculinizing elements in engineering teaching environments. The elements they identify contribute to difficult organizational cultures and structures in higher engineering education and may, in part, explain lower rates of persistence for women engineers. A trend of "male dominance" was found in perceptions of engineering in society. The engineering field was classified as having a "masculine image". European female students characterized it as: "machine-oriented, with less communication, rational but not creative, not positive, but combined with high earnings" (Sagebiel and Dahmen 2006, 9). Coping strategies such as addressing feelings or attitudes towards gender discrimination, welcome events and activities, interdisciplinary tracks, and single-sex courses employing a more women-friendly inclusive culture were suggested for the field of engineering. In another example, a European project, Curriculum Women and Technology, CuWaT (1998), provided evidence for elements effective in the retention of women in engineering, through cross-disciplinary courses, group work and project work (Sagebiel and Dahmen 2006). Beraud (2003) extended this by highlighting, through his research on Engineering in Europe, that interdisciplinary degrees appear to be more attractive to female students than single-track engineering degrees and that more women could be recruited to engineering by utilizing this interdisciplinary factor. The lower participation of women in Europe persists despite these and other efforts, despite growth in general educational attainment for women, and despite reports of high value for female engineers in industry and academia presently and as integral to the future global economy (Beraud 2003; Brush 1991).

Policymakers have tried to respond by applying best practices that have been identified as important for women's participation in order to support women in engineering. Thanks to global norms and programs stressing increased educational access for all students, with particular focus on the increased educational access of girls and women, opportunity has spread through the direct influence of international and national organizations such as UNESCO and the World Bank (Bradley and Ramirez 1996). Some areas of women's participation (e.g., general higher education enrollment) have seen notable progress towards gender parity (Ramirez and Wotipka 2001). Since this growth has not universally affected engineering fields, national and international governmental organizations in North American and Europe have mobilized billions of dollars to identify and then address the underlying factors that have prevented gender progress in all areas and have implemented strategies to increase female participation as students and professionals within scientific fields. See, for example, the US National Science Foundation and the US National Academies and the Central European Centre for Women and Youth in Science. Even so, the efforts of policymakers in North American and Europe have had relatively minimal impact on overall numbers (Zoli et al. 2008).

As we note above, the characteristics of women's participation in engineering vary widely between national contexts. Although growth in women's participation has stagnated in US and the UK, it has expanded in other places. According to UNESCO data for 2012/2014, five countries accounted for the highest gender discrepancy towards male-dominated engineering, manufacturing, and construction management programs: Germany (81%), Japan (88%), Saudi Arabia (94%), Switzerland (84%), and the U.S. (83%) (UIS 2015). As we note later in this chapter in the Saudi Arabia country brief, the most probable reason for the low numbers of women enrolled in the kingdom is because the first undergraduate engineering program for women was opened in 2012 at the public King Abdulaziz University, and the graduate-level programs offered at KAUST since 2009 are for the international and national elite, and thus not representative. In contrast, in Saudi Arabia, women constituted 72% of graduates in scientific fields in 2011 (Matthews 2013). Females in science in the five GCC countries for which UNESCO published data in 2015 were strongly represented: Jordan (38%, 2012), Kuwait (43%, 2012), Oman (37%, 2013 up from 25 % in 2011), Qatar (34%,

2013), and the UAE (35%, 2013). In Jordan, in addition, 49% of computer science undergraduates in 2011 were women, according to government statistics (Matthews 2013). Mongolia, Greece, Serbia, Panama, Denmark, Bulgaria, and Malaysia constitute a diverse group of countries where women represent a third of all engineering graduates (Charles 2011). Although Charles and Bradley (2009) found that countries with high GDPPC had greater gender typing of curricular fields (e.g., engineering = male or nursing = female) than those with low GDPPC, Abu-Lail et al. (2012) and Lagesen (2008) found in their interviews of women who chose engineering as a field of study that Jordanians and Malaysians may gender-type subfields of engineering rather than engineering as a whole. In their trend research, Ramirez and Wotipka (2001) found some minor irregularities in science and engineering globally and that women more often chose science over engineering.

To further complicate the variation between countries in women's participation and the mechanisms explaining imbalances, it is important to note that the issue does not exist in a vacuum in isolation for each country. Differences in opportunities for women engineers may also translate to patterns of international skilled migration in engineering fields. Immigration and emigration statistics of foreign-born individuals with science and engineering degrees reveal high proportions of skilled workers in engineering disciplines and science/engineering overall who were born abroad and are now working in the US, as one example (Regets 2007). The share of women in the US's immigration flow was over half (55%) in both 2000 and 2010, and women make up an increasing proportion of highly skilled migrants (OECD 2012). Interestingly, international migrants in recent years are less likely to settle abroad permanently (Iredale 2005). The diverse, interwoven, and developing picture of women in engineering demands that we more deeply understand the context and trends within countries and the way in which these contexts might inform one another.

Country Briefs

We now take a deeper look at women in engineering in one example GCC country as well as a set of comparison countries in the MENA region, Southeast Asia, and North America. These country briefs provide implications for future institutional development in the GCC and, lay the groundwork for further study of women's participation across national contexts. First we highlight Saudi Arabia, one of the GCC countries we have already discussed in detail. To complement our detailed GCC information, we include a brief on another country in the Middle East (Jordan), one in North Africa (Tunisia), another majority-Muslim country in Southeast Asia (Malaysia), and a high-GPDDC country where much research has been done with little change (US). Table 2 provides broad general summaries of the information that we detail in our country briefs below.

Table 2. Country Profile Data Descriptions

	Saudi Arabia	Jordan	Tunisia	Malaysia	US
Education	Higher female than male enrollment rate in tertiary overall; high proportion of females attaining degrees	Higher female than male enrollment rate in tertiary overall	High female secondary enrollment (89%), lower tertiary enrollment (21%)	High female literacy rate, minimal literacy gap by sex	High female enrollment, attainment overall; lower engineering and STEM
Economic	Low female employment rate (20.1%); women likely employed in administrative/ academic positions, public sector; low female STEM researcher rate	Low but growing proportion of women in engineering and computer science, higher in other sciences	Women represented in engineering, often in public or academic work	High proportion of female STEM researchers (48.7%)	High female employment; major dropoff in female engineering/ STEM participation during employment
Cultural, historical, political factors	Gender roles and gender segregated schooling adapting to new institutions and new economic demands	Gender typing of subdisciplines but not whole area of STEM	Historic support for women's equality; national union for women	Gender typing of subdisciplines but not whole area of STEM; acceptance of women in engineering industry, even in managerial roles; continued expectance of traditional family roles	High income democracy, multiple women's movements, policy focus on women's participation and women in engineering/ STEM
Available research	Some information from management and business literature, knowledge economy literature	Little available research on this question	Some literature from economics/ World Bank programs, literature from sociology	Little available research on this question	Great deal, from basic participation, documentation of women in engineering programs, to rigorous research

Source(s): UIS Statistics, 2015.

As we show in Figure 1, there is a great deal of heterogeneity in the relationship between female representation and national income level (and certainly other national characteristics). We highlight the four countries for which we provide country briefs that Charles and Bradley (2009) include in their study. We also approximate the placement of the six GCC countries on the same graph. Interestingly, five of the six GCC countries have higher representation of women in engineering than their GDP per capita would predict. The exception is Saudi Arabia, which has a much lower proportion of university graduates who are women in programs in Engineering, Manufacturing, and Construction, which is the proxy measure we have available for women's representation (3.44%; UIS 2015). However, the proportion of graduates who are women from science and technology programs (38.9%) and science programs (57.2%) is as high as or higher than other GCC countries (UIS 2015), and we delve into the Saudi case later in this section.

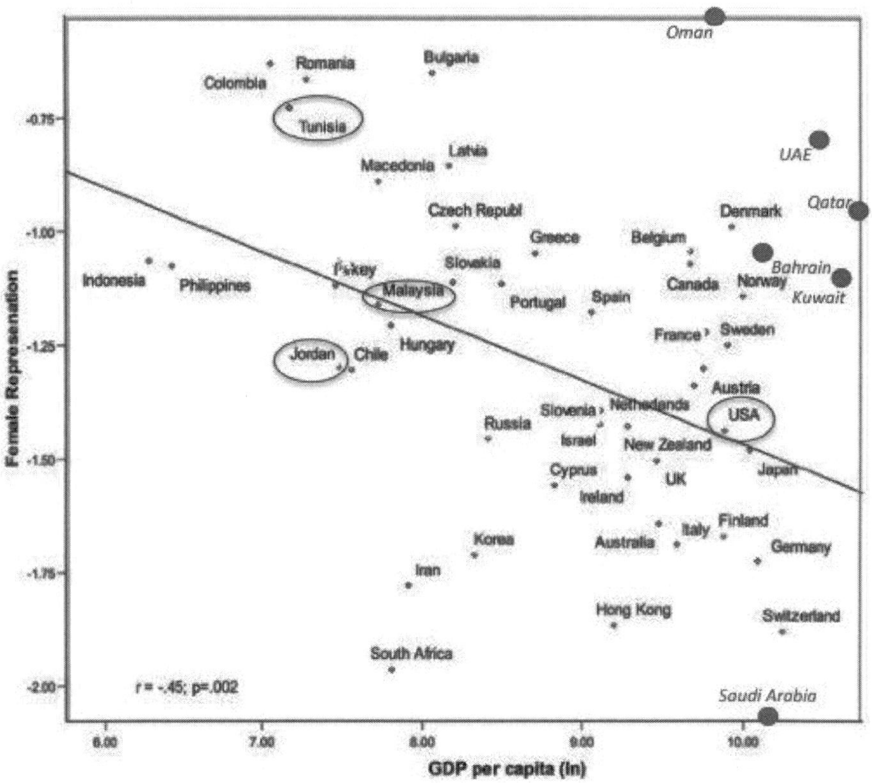

Figure 1: Country-level Female Representation in Engineering by GDP per Capita (adapted from Charles and Bradley [2009]; UIS, 2015) N.B. In our additions, the female representation of Oman is higher than the range given in Charles and Bradley [2009], and that of Saudi Arabia is lower. Also, the logged GDPPC of Qatar extends beyond their range.

Table 3 summarizes key metrics that we highlight in each of the country briefs that follow. It does not list the US metrics, as we delve into other literature beyond the UIS figures in the US section, based on the greater amount of literature available.

Table 3. Literacy Rates, Tertiary Enrollment & STEM Researchers

	Youth Literacy Rate		Tertiary Enrollment		Literacy Rate		STEM Researchers
Country / Percentage (%)	**F**	**M**	**F**	**M**	**F**	**M**	**F**
Saudi Arabia	97.0	99.0	46.0	41.0	-	-	1.4
Jordan	94-95	99.0	43.0	37.0	-	-	22.8
Tunisia	96.0	98.0	21.0	25.0	71.0	87.0	-
Malaysia	99.0	98.0	-	-	91.0	95.0	48.7

Source(s): UIS Statistics, 2015.

Saudi Arabia: High Income, High Female Attainment, High Female Unemployment, Gender-segregated Education

Saudi Arabia has shown growth over the last few decades in women's participation, but this is particular to the public sector and to jobs in education. The country's economy is healthy, with a notably higher GDP per capita than its MENA counterparts of Jordan and Tunisia. Saudi experiences near parity in the youth literacy rate (97% /99% female/ male 15-24 years of age) and higher female tertiary enrollment (46% vs. 41% female/male; UIS 2015). However, the seeming gender-parity within the educational system does not translate into equality in the labor force; in STEM, only 1.4% of researchers are women, and overall female employment hovered around 20.1 % in 2013 as compared to the GCC member average of 43%, the lowest in the region (UIS 2015; G20 Labour Markets 2014; Islam 2014). At present, only 14.4% of women are represented in the Saudi workforce, and as many as 78.3% of unemployed women are university graduates or degree-holders. The government hires women predominantly in education (85%), and, as of 2014, women cover only 5% of the jobs in the private sector. Statistics on female participation in the workforce show that only 26.0% of Saudi females are employed in a technical field, while 96.0% are in educational roles, and 90.0% are in administrative positions. Thus, administrative, academic, and medical areas dominate female employment in the public sector; whereas the technical and vocational areas house private sector female employees (Calvert and Al-Shetaiwi 2002).

Al-Mubaraki (2011) discusses how higher education in the Kingdom of Saudi Arabia (KSA) has been growing rapidly, with large-scale changes and fast-paced progress and growth. Global movements in higher education, requirements for an increasingly specialized national workforce, and the possibilities and difficulties associated with both areas (academics and employment needs), together have effectively led towards the formation of present guidelines and future strategies. This has given birth to the "Saudi model of a knowledge-based economy" (Al-Mubaraki 2011). Onsman (2011) discusses the fact that KSA's higher education development holds the primary concern of upholding and preserving its strong socio-culturally Arabian foundation, while simultaneously working hard towards gaining international significance (p. 519). Islam (2014) reports that 58% of Saudi students at the university level are women, and a scholarship program for women has been put forward by the Saudi government that is one of the largest of its kind globally.

The picture of women's participation in the Kingdom had already started to shift at the turn of the century. The number of women in Technical and Vocational Education/ Training (TVET) increased sharply between 1973 and 1999 (Calvert and Al-Shetaiwi 2002). This was in major part due to the government's encouragement of and emphasis on the value of TVET for women, bringing some change in societal attitudes towards women's participation in STEM areas. Recognizing the high numbers of female university graduates and their low employment numbers as a national issue of underutilized human resources, the Saudi Ministry of Labor began to implement a series of policies and programs in 2011 that were designed to increase the number of women in the workforce, which continue today (G20 Labour Markets 2014). According to Saudi Arabia's G20 2014 Employment Plan, 8.8 million of the 12.4 million female nationals are of working age. A significant partnership between the Saudi government and Harvard's Kennedy School of Government began in 2014 in order to address the myriad technical and societal challenges that ensued after policies designed to usher women into the workplace went into effect (Searcey 2014).

Gender segregated schools are viewed as a possible vehicle for the education of the nearly "… '80% of girls in Saudi Arabia…interested in engineering' [according to a recruitment expert in the Middle East]" (Mafeo 2013). The ubiquity of gender-segregated schools is seen by some as promotion of traditional women's expectations and roles (Baki 2004). Educational institutions themselves are perceived as a mechanism of control and enforcer of the traditional and lower status of women (Doumato 2003). Secondary public school curricula for girls is cited as being up to 20 years out of date, with the topical focus being primarily on religious studies, Arabic Language, and home economics (Al-Munajjed 2009). Women continue to be under-enrolled in technical and vocational education (TVET), where jobs are reportedly undersupplied (Calvert and Al-Shetaiwi 2002). And, up until very recently, the field of engineering had excluded women from studies (Hamdan 2005).

This, too, is changing. In 2012, King Abdulaziz University in Jeddah opened the first college of engineering open to women in public universities, offering degrees in two areas: electrical and computing engineering and industrial engineering. Given the overrepresentation of women in the public sector, Calvert and Al-Shetaiwi (2002) propose a push towards increased Saudi female representation in private sector employment, and they suggest factors that would contribute to this policy goal. They suggest more cooperation between public and private sectors in the planning of TVET and establishing of direct links between women's education and the labor market. Their research implies that progress could be made by altering the structure of technical and vocational education in Saudi Arabia. Quamar (2013) states that over the past two decades there has been a rapid upward shift in quality and availability of education for Saudi women. These improvements have naturally led to increased numbers of educated women and subsequent positive growth in all related fields, such as socio-political and economic status.

Co-educational opportunities do exist, but they are rare and are not part of the Kingdom's public school system. The sole higher education example is King Abdullah University of Science and Technology (KAUST), which has offered graduate-level programs (Master's and PhD) in engineering since 2009 to both Saudi and international women and men. KAUST's enrollment of Saudis is capped at 50% in order to make the student body truly international; Saudi enrollment is currently at 35% (Havergal 2015). The university also includes "research funding for global partnerships and collaborative research with the private sector to advance economic development" (Ulaby 2010, 178). KAUST has proactively increased its female representation in both the faculty and student body, with women comprising 35% of both sectors (Havergal 2015). In the 2014-2015 academic year, the university was hosting 840 students from 69 countries, 302 of whom are women and 246 of whom are Saudi Arabian (Butler 2015). KAUST is an elite institution and, thus, does not impact a significant percentage of the Saudi female cohort in engineering programs or in the workforce. There has been a notable amount of policy literature devoted to increasing women's participation and role in the Saudi economy; yet, little is known about how the overall system impacts women's decisions to enter engineering or computing or to remain in it as a profession.

Jordan: Recent Growth in Representation, Less Research Focused on this Issue

Jordan provides an interesting complement and counterexample, another Middle Eastern country, but outside the GCC. Jordan, with a GDP per capita very similar to another MENA country we study, Tunisia, has seen a greater historical change in women's participation in engineering over its history, as the participation of women began at a much lower proportion but has increased in recent years (IGU/UNESCO 2012). In Jordan, women's and men's literacy rates are close to equal (94% /95% for women and men overall and 99% for female/male ages 15-24; UIS 2015). Additionally, the country enjoys near parity in secondary educational enrollment and notable overrepresentation of women in tertiary enrollment (43%/37% female/male enrollment; UIS 2015). This gender parity in enrollment is a recent phenomenon, and it has raised the question of a possible change in the perception of young women's roles in society (Adely 2004). Still, fewer than a quarter (22.8%) of STEM researchers are women (UIS 2015). Although the percentage of female students in specific disciplines such as computer science is lower than men (about 25% in 2009 and, as cited above, about 49% in 2011), it is rising and approaching parity (Khreisat 2009); further, the proportion of women in other STEM disciplines (e.g., chemistry, biology, physics, and math) may reach as high as 75%. Scholars have yet to identify what factors have led to such a different and rapidly changing outlook for women in engineering in Jordan.

Tunisia: Historical Female Participation, Secularism, and French Education

Tunisia provides a distinct contrasting example to the GCC and Middle East context. Historically, Tunisian national policymakers have explicitly embraced equality for women (Murphey 2003). However, recent studies found that Tunisian engineers from multiple sectors noted that traditional gender perceptions and expectations tied to familial roles did not fit with women's participation in engineering work (DeBoer 2007; Zghal 2006).

The traditional, historically French education system in Tunisia is seen as having a leveling effect on educational opportunity, and overall educational enrollment figures are approximately equal for women and men, tipping slightly towards women in the secondary (89% /93% for boys and girls, respectively) and tertiary (25% /21% for women and men) levels (UIS 2015). Researchers in STEM fields are nearly half women (47.4%; UIS 2015). The female-male literacy gap, which is notable in the overall population (71% /87% female/male), is reduced in youth (96% /98% female/male in 15-24 year olds; UIS 2015). Tunisia's marked secularism, its active civil institutions, and the longstanding involvement of groups like the National Tunisian Women's Union have improved women's engineering participation. Some sociological work has been done to understand the current experiences of female engineers and to understand the development of the Tunisian engineering education system (Zghal 2006; DeBoer 2016).

Malaysia: Near Parity in Academia, Perceived Need to Reconcile Gender Roles

We turn to Malaysia for the perspective of another majority Muslim nation, here located in the distinct geographic region of Southeast Asia. Malaysia's historically British education system has been combined with state efforts at modernization and support of global competitiveness. Malaysia has a high and relatively equal literacy rate (91% /95% female/male for the population, and 99% /98% for female/male ages 15-24), and over-representation of women in secondary and tertiary enrollment (UIS 2015). Indeed, nearly half (48.7%) of STEM researchers are women (UIS 2015). Despite this, executive-level female employees in a Malaysian multinational company indicated that family structure and women's commitment to family were the most significant barriers to women's career progression (Ismail and Ibrahim 2008). These findings corroborate other analyses of Malaysian women in managerial roles and of Malaysian computer scientists, wherein a specific, well-understood construct of female identity had to be reconciled with academic career aspirations (Lagesen 2008; Luke 2002; Mellström 2009).

The United States: Huge National Investment Resulting in a Plethora of Research, Yet no Real Increase in Numbers

In the US in the 1950's, women represented less than 5% of the graduating classes in schools of law, medicine, and engineering (Diekman et al. 2010). During the intervening years, despite overt and covert discrimination, US women fought in both the courts and public opinion forums to be admitted into schools of law and medicine (both human and veterinary). Women now comprise 50% or more of the graduating classes in these professions (Kam 2005) and are overrepresented in higher education in the US as a whole (Shavit, Arum, and Gamoran 2007). Many claim this majority portends a future decline in gender segregation in fields traditionally dominated by men, as well as an overall decline of gender separation in all areas of the public sphere (Charles and Bradley 2002). Yet, recent studies show that gender segregation is unrelentingly high in the US, and particularly in engineering, as it is in other high-income countries in North America and Western Europe, where legal and societal structures to encourage gender equality in educational and public spheres have been in place for a number of decades (Charles and Bradley 2002). Thus, the same progress in fields traditionally dominated by men, such as law and medicine, has not been matched in engineering, where the proportion of women in the field has remained around 11% for the last 20 years (Fouad and Singh 2011).

Studies on the underrepresentation of women in engineering in the US are abundant. Since the 1970's, the US federal government, industry and professional engineering societies have contributed billions of dollars to increase the number of women in US engineering programs with minimal impact. These funds have been used to 1) study and understand the explicit and implicit patterns of discrimination against women in engineering schools and workplaces, 2) identify and campaign against incorrect and negative stereotypes about women's cognitive abilities with respect to science and engineering, 3) attract young women to the engineering profession by convincing them that engineers make a positive impact on society, and 4) retain women once they are in engineering degree programs by providing (among others) alternative instruction, mentors, and role models (National Science Foundation 2011).

These efforts have helped to increase the women graduating from undergraduate engineering programs from 2% in the 1970s to 17% in the 1990's to 20% in 2004 and 19% in 2015 (American Society for Engineering Education 2015). According to a 2010 study, women in the US were found to be less likely to perceive that STEM (especially engineering) careers would fulfill communal goals, a belief that was related to women's lower likelihood of choosing engineering as a field of study (Diekman et al. 2010). The 2011 report, *Stemming the Tide: Why Women Leave Engineering*, highlighted challenges for female engineering graduates in the US (Fouad and Singh 2011). That report indicated that 20% of women left the field because of working conditions (lack of advancement opportunities and low salaries were among reasons cited), unwelcoming workplace climate and culture, and the desire to spend more time with family. The study also revealed that one third of women never entered the engineering workplace because they perceived engineering as an inflexible career or because the engineering workplace culture did not support women. Although the consideration of women navigating perceptions of appropriate gender roles at home and in the workplace runs across countries, more differences than similarities emerge. The notable difference in participation—stagnant growth and lower participation—in the US stands out, though. It is for this reason that our research group has embarked on an investigation of systematic differences in women engineer's experiences across countries.

Online/Digital Education as an Accessible Avenue for Education

The statistics we have discussed and the research questions our team has developed frame the current systems in place, systems that differentially facilitate women's participation in engineering in the Gulf region. In this penultimate section, we describe novel avenues for educational participation and provocative, more contemporary areas of growth. Online education has rapidly expanded, potentially fostering opportunities for women, youth, and rural citizens. Indeed, numerous GCC online learning or "e-learning" initiatives were created in order to support post-oil knowledge economies (Weber 2010). We highlight one particular recent example from Saudi Arabia. New platforms for online and blended education may indeed support the growth and variation in women's engineering study and practice. Massive Open Online Courses (MOOCs) are one such platform. Policymakers must be cautious about potential outcomes of their use of new educational structures, especially when unintended consequences are not well understood.

Online/digital education as an accessible avenue for education

Beginning at the end of the last decade (2000-2010), efforts to broadly scale online learning met with technological growth in the capacity for online platforms to connect and automate the grading of thousands of students' work. In 2011-2012, the first broadly disseminated MOOCs were presented from Stanford and MIT. Hundreds of thousands of students signed up and virtually matriculated into classes that included online lecture videos, auto-grading of homework problems, and massive discussion forums. This included notable participation from virtual "students" across Gulf Cooperation Council countries, in particular in engineering and computer science content-related courses (Breslow et al. 2013; Ho et al. 2015).

Although absolute numbers in terms of participation were high, the question of whether these courses were achieving their mission of rapidly expanding access to high-quality education where it was previously unavailable was questionable. A number of investigations of student backgrounds and performance profiles found that MOOCs might not have been truly increasing access (DeBoer and Breslow 2014a; DeBoer and Stump 2014), as many students already had university degrees, usually in the subject they were studying on the MOOC. One major barrier to differential access was likely language, as early courses were largely offered from English-language universities in Europe and North America (DeBoer and Breslow 2014a; DeBoer and Breslow 2014b)

Recognizing language as a key barrier to policy diffusion (DeBoer 2012b; DeBoer 2012a), to cross-border partnerships, and to participation in MOOCs as a global technology, edX forged a partnership with Jordan's Queen Rania Foundation to create a MOOC portal focused on the Arab world (O'Connell 2013). The platform, *Edraak*, would provide Arabic translations of existing edX courses chosen for relevance to a range of issues. In addition, plans were laid for *Edraak* to develop its own courses in Arabic in collaboration with prominent Arab faculty and professionals. (Similar collaborations had been developed, for example, with Chinese colleagues, creating "XuetangX".) The not-for-profit platform launched in 2014, featuring courses from edX founding partners Harvard and MIT as well as "the best Arab instructors [and] regional academic institutions" (Agarwal 2014). *Edraak* explicitly targeted its efforts towards the large Arabic population of youth. According to Queen Rania, as of March 2015, the platform had grown from 10 courses and 85,000 users to 20 courses and over 140,000 subscribers (Jordan News Agency 2015). The *Edraak* platform has provided access to Arabic-language courses not only for Jordanian students, but for students across the peninsula, including students in Syria, where the security situation has prevented their attendance at university; the Jordanian queen referred to the role of technology in purportedly increasing access in the Arab World as the "Startup Spring" (Gordts 2015).

After its 2013 announcement (O'Connell 2013) of collaboration with Jordan to create an Arabic-language specific platform, edX more recently forged a collaboration with Saudi Arabia (edX 2014). The partnership was made specifically between edX and Saudi's Ministry of Labor (specifically, the Ministry of Labor's Human Resources Development Fund), with an eye towards increasing access to job opportunities and more closely linking potential labor force participants to growing private sector needs for skilled work. Pointedly, the announcement targets bridging the gap between education and employment for women (as well as youth, the disabled, and rural citizens) (edX 2014). This platform, launched in December 2014 as *Doroob*, explicitly targets job skills and underemployment (Taibah 2015). The Saudi Advisory leading the initiative cites as justification for these efforts the fact that "50% of YouTube streamers in Saudi Arabia are women and 40% of content they view is education" (Taibah 2015).

ICT Opportunities Constructed Explicitly for Women

The Saudi platform's explicit focus on women (as well other populations that have less access to higher education) highlights the technological "fix" offered by distance education in general and the recent advent of MOOCs in particular. Women in Saudi Arabia report increasingly wanting to work, but they face barriers to participation such as transportation to the job site (Searcey 2014). Online education supports advanced learning that can happen at home, addressing both women's individual aspirations and a broader socioeconomic need without necessarily addressing prevalent social norms and practices. Indeed, online learning may address the particular work patterns of women, who, in Saudi Arabia and other emerging markets, more often participate in the informal economy (G20 Labour Markets 2015).

Some in the popular press (Gais 2014) and on the ground (Searcey 2014) perceive MOOC platforms and such alternative avenues to access as band-aid fixes for two reasons. First, the provision of increased education for women may not be as pressing a need as initially presented. There is already a notable mismatch between degree attainment and work profiles for women, who are on average better schooled than men; indeed, women make up more than half of university graduates in Saudi Arabia (Gais 2014; Searcey 2014). This means that MOOC platforms may again be delivering education to a population that already has access to higher education opportunities, whose needs are not so much for educational attainment as for labor force opportunity. Interestingly, MOOCs may unintentionally provide a more useful opportunity given the prevalence of ECS courses in their offerings. One of Saudi Arabia's major labor force needs is in the sector of information technology, where jobs may involve remote work, and women in these positions may still adhere to more conservative social norms.

However, critics see another mismatch in what MOOCs offer and what is needed in Saudi Arabia. As has happened in previous educational reforms, information and communications technology (ICT) may again serve to preserve women's distance from normative educational opportunities (e.g., such strategies were used before in schools to separate female students from male lecturers, Gais 2014). Distance education and MOOCs may again preserve sex segregation in education and the labor force. Although this segregation may reflect women's desires and the utility they perceive in e-learning initiatives (Weber 2010), this practice may create additional barriers to participation in multi-national STI partnerships and hamper the development of the Gulf's knowledge economies to their full potential.

Aswad, Vidican, and Samulewicz (2011) acknowledge, through their study, the importance of understanding how the local socio-cultural contexts affect individuals and the economy and of altering policies to better fit local constructs. They stress upon the need for rising above negative connotations of engineering and increasing awareness of content information and expectations, as well as related career opportunities that are culturally acceptable, in order to increase participation and interest in STEM fields. Syed (2011) investigated and put forward strategies for producing higher rates of success for female entrepreneurs in Islamic countries, where such activities hold great economic potential and transformative power, when utilized and channeled appropriately. Metcalfe (2008) further revealed significant progress made in female leadership and socio-political empowerment, with some ongoing institutional and cultural barriers embedded within systems and complexities local to the Middle East.

Women as an Engine of Growth

As we note, women have already played a significant role in supporting the rapid development of GCC economies. Beyond female enrollment in engineering in higher education and attending universities, there is evidence that the participation of women in the technical innovation enterprise has been an important component of growth in the GCC. Policymakers have recognized this and begun to create recruitment programs. Watson et al. (2006) have outlined various programs dedicated to recruiting and retaining females in the areas of education and employment in engineering.

Naser, Mohammed, and Nuseibeh (2009) researched the factors affecting female entrepreneurship in the region and in the UAE in particular, with a focus on what motivates women to establish their own businesses. Findings showed the importance of governmental financial support (initial capital), and links to family businesses (husband/father), along with the presence of self-sufficient feelings, knowledge, skills, and experience. However, social norms, competition, and market networks did not prove to be obstacles in the development of female entrepreneurs. Women from across the Middle East are finding the United Arab Emirates a business friendly environment for developing their own professional ventures. Many of these entrepreneurs left established positions to start their own businesses focused on effectively addressing issues affecting women in what have been predominately male-dominated sectors in the region, such as computing and IT, professional services, and news (MacKenzie 2014; Syeed and Zafar 2014).

Worldwide, only 10% of tech entrepreneurs are women; in the GCC, women have 35% of the market share (Hosea 2014). The following three examples represent some of the best-known female tech initiatives in the Gulf region. Baraka Ventures was founded in 2006 to fund regional socio-technical enterprises that benefit the community in a sustainable way; one of the 2012 investments resulted in the on-line news agency Barakabits "Good News from the Middle East" to capture and communicate the vibrant stories emerging from the region. Nabbesh, a recruitment website founded in 2012 "to address the Arab World's unemployment challenges by virtually connecting expert regional talent with flexible work, online jobs and opportunities to work from home," claims approximately 35% of their clients as women in the GCC (Hosea 2014). The Arab Women in Computing organization whose goal is "to support, inspire, retain, and encourage collaboration among, increase visibility of and help elevate the status of Arab women in computing, and allow them to achieve their career goals" was established in 2012 (ArabWIC 2014).

Some say that the internet is a more meritocratic space for women to excel than other professional environments in the region and that Arab women living in the GCC are taking advantage of that opportunity (Hosea 2014). While these are three of many success stories for women in the computing and IT fields in the UAE, little is known about the core reasons women pursue engineering and computing fields in the UAE, why the UAE is considered a female-friendly environment for entrepreneurs, and why Emirati women with tertiary degrees in engineering and computing continue to not enter the workforce at rates that mirror the degrees awarded. Another country in our study, Jordan, has similar participation by women and similarly little research on the core reasons for why women choose engineering or computing as a field of study or profession.

Lessons, Implications, and Recommendations

Our international comparative study of women's participation in engineering education and in engineering work demanded that we investigate the state of STEM and women in the Gulf overall, as well as corresponding contexts in the MENA region and globally. As we have shown in our discussion, our presentation of important background information from the GCC, and our contextualization of GCC data within a regional and global frame, the state of women in engineering and computing in the Gulf is a diverse picture, full of potential, and, in many places, trending upwards. The diversity of successes and challenges faced by the region on women's inclusion in STI make generalized policies or consequences rare, but we observe a few key emerging themes in our work. We provide here four implications that resonate with each of the complex nation states of the GCC as our concluding section.

New Policy Opportunities Must Take Into Account Potential Outcomes Both for Near-term Changes as well as for Far-term Access and Equity

Although there are early successes from some policies that have increased women's access, new avenues for education and programs to increase women's participation must be done with planning beyond temporary "quick fixes". For example, single-sex schools are common throughout the GCC, and these may be an important component of the current growth in women's degree attainment. Zeid and El-Bahey (2011) present a case study in Kuwait on the effects of introducing single-gender classrooms in a software engineering course. Findings show an upward trend in academic performance as a result of segregation, as well as a significant decrease in the performance gap between the two genders. They propose this solution as a useful strategy for improving learning in higher education and reducing problems of gender divide in the workplace. Boashash (2013) explains through a comparative analysis that engineering education in the GCC/MENA region needs to be attuned to local customs – mindful of cultural and religious norms, based on local research, with contextual adaptations – in order for engineering education to hold its own on the international platforms. Single-sex schooling may provide such a tailored response.

However, continued educational segregation may also delay the inevitable, as women entering the workforce will need to work alongside men, whether in the GCC or in multinational institutions. MOOCs and the application of online learning pose challenges for long term success in the same way as single-sex schools—women will, at some point in their trajectories, be part of an integrated labor force. And, their increased participation throughout their educational trajectories and into the labor force will buttress GCC goals for active knowledge economies. This means integration into collaborative classroom environments and preparation for global opportunities. One concrete policy recommendation is to support increased numbers of female teachers in STI content areas. This follows on extensive research supporting role models and mentors as key to long-term retention for both women and other under-represented groups in STEM fields (Carrington, Tymms, and Merrell 2008). In Kuwait, for example, high proportions of female teachers in primary education (over 90%) are strongly related to a high female gross enrollment ratio in secondary education (approximately 90%) (UNESCO 2012). Building sustainable solutions into educational structures will foster long-term women's participation in the Gulf and better serve institutional and cross-institutional goals.

As another example, GCC states have rapidly increased the number of higher education institutions locally, often by incorporating foreign branches of Western universities and/or recruiting international faculty. As the call for the 2015 Gulf Research Meeting states,

> The GCC states have made major advances over the last half century in widening access to post-secondary education. For example, eight universities were operating in Saudi Arabia in 2003. Since then at least 100 additional universities and colleges have opened, and the country's annual budget for higher education has reached $15 billion, for 23 million inhabitants. In addition, the United Arab Emirates (UAE) and Qatar have authorized 40 foreign branches of Western universities, most of which date only from the last decade. (Abusharaf and Eickelman 2015)

Although this strategy has dramatically expanded the state of higher education in the GCC, it has not kept in mind the long-term local capacity development of local faculty and institutions. Many of these universities currently have very few native students and even fewer native faculty. For example, at NYU Abu Dhabi, about 10% of the student body is Emirati of the 140 students in their first graduating class in 2014 (Al Bustani, 2015). By contrast, at Carnegie Mellon University (CMU) Qatar, about 40% of the student body is Qatari, but overall numbers are still low (about 400 matriculated in 2014; CMU Qatar, 2014). Although these prestigious collaborations may enrich the host country in a variety of ways, the number of native students impacted is relatively low. Even the benefits for the host country have been called into question, and the universities are difficult to gain admission to and considered elite by local students. They are, therefore, not currently sustainable pathways by which to substantially increase the number of women in engineering and STEM in the GCC.

Building a Competitive Knowledge Economy is a Major Goal of the GCC as a Region and as Individual Nation States

GCC countries are reshaping their tertiary education and STI workforce profiles in order to create regionally and globally competitive knowledge economies. In order to fully realize their potential, however, the *knowledge economy* literature stresses the importance of women's involvement. For example, a report from the Gender and Development Group of the World Bank begins:

> Gender equality is a legitimate policy goal in and of itself...The evidence presented in this paper, however, suggests gender equality is also desirable from an efficiency perspective: increases in opportunities for women lead to improvements in human development outcomes, poverty reduction, and—although evidence on this last point is relatively weak—potentially accelerated rates of economic growth. (Morrison et al. 2007, 1)

Each of the six GCC countries is committed to building a knowledge economy. Many of the existing universities have partnered substantially with universities beyond the GCC, ranging from Carnegie-Mellon University in Qatar to NYU Abu Dhabi, to others we have not mentioned in our chapter. The cross-continent partnerships have actively engaged institutional partners in Western Universities. As we describe above, though, many of these partnerships have not yet yielded major impacts on enrolling local students. One potential avenue for these cross-continent collaborations to influence women's participation and the growth of local knowledge economies is through their support for local women students and faculty. The advent of CMU Qatar, NYU Abu Dhabi, and such institutions has required consideration of new policies for female students and faculty, and such policy levers could concretely impact policy changes at other local universities and provide notable opportunities for local women engineers.

Further, in order to build a knowledge economy that capitalizes on universities and effectively leverages their connections, GCC institutions need to explore partnerships within the GCC and the MENA to build regional capacity by leveraging expertise and to identify common challenges and join forces to address them. These partnerships can and should involve leading female students and faculty from the region, as growing global economies must take into account shifting gender identities and roles (Gillard et al. 2008). Where institutions and societal gender norms previously created systematic barriers to women's inclusion, shifting demands for knowledge economies and even specifically engineering, computing, and information systems require re-scoping avenues for women in education and STI practice. Policy levers, such as those used in Qatar to expand opportunities for women, must be transformative and targeted; indeed, in Qatar, the changes in policies, education, and the social sphere have encouraged women to tend to persist in their professions even through child-bearing years (Felder and Vuollo 2008).

Researchers and Policymakers Must Recognize and Understand the Diversity of Contexts Within the GCC, Which Demands Tailored Policies and Even Tailored Research Questions

Our ongoing empirical research intends to unearth the "relevant" questions to ask in each national study site regarding women's education and work in engineering. Each GCC context (as well as the other non-GCC countries in our study) have different histories, economies, social and cultural norms, and current profiles of women in STI. The situations vary between countries for a variety of reasons. The GCC's national histories lead to different explanations for women's participation and necessitate relevant research questions; although we do observe some common GCC trends, both study questions and policy solutions must be tailored to the local context. We have found that only through an iterative process of research question generation and refinement, alongside a local interdisciplinary team, can relevant research strategies be created. Even then, we find that these research questions may actually apply to different contexts as well, albeit in different ways.

More Data are Needed as Women's Participation in STI Expands and Diversifies

Data quality, availability, and relevance remain an issue, especially as policymakers try to understand potential effects of novel reforms and decide on best next steps. More detailed information on women's STI participation would add necessary nuance: for example, by specific engineering discipline (mechanical engineering, electrical engineering, etc.), by institution, by subnational division (e.g., local region, city, urbanicity), or other key demographic factors. Efforts by the UNESCO Institute for Statistics and the International Labour Organization provide meaningful aggregate numbers, but even these national statistics have sporadic years of data missing, may not reflect a fully representative sampling frame, or combine engineering with fields such as construction, which may greatly obscure trends of interest.

Conclusion

More information utilized to inform the "right" relevant questions will foster productive long-term-oriented policies supporting women's full participation and contribute to integrated knowledge economies across the GCC. Increased regional and international collaborations between higher education institutions are certainly important to nurturing the growth of women's achievements in STI, and may actually help to drive broader women's access. Women in the GCC are already participating in engineering and computing in a variety of ways, in some areas, at greater rates than in the United States. Women's networks within GCC countries and throughout the region may have policy and programmatic lessons learned that could nurture women's access in other areas of the globe. Historically, changes in the labor force drive changes in societal norms. The considerable and growing presence of women in the GCC's IT and broader engineering labor force may, in turn, serve to change broader and ingrained perceptions and practices of gender roles and identities. Women are already participating in engineering and computing in GCC countries in a variety of ways, and they should be empowered to support each other, to support the next generation of female engineers, and to accelerate the growth of the burgeoning and diversifying STI enterprise in the GCC.

References

Abu Lail, Nehal I., Phang Abdullah, Fatin Aliah, Ashley Ater Kranov, Khairiyah Mohd-Yusof, Robert G. Olsen, Rochelle Letrice Williams, and Azizan Zainal Abidin. 2012. "Persistent Gender Inequity in US Undergraduate Engineering: Looking to Jordan and Malaysia for Factors to their Success in Achieving Gender Parity." *Proceedings for the 2012 American Society for Engineering Education Annual Conference.*

Abusharaf, Rogaia Mustafa and Dale F. Eickelman. 2015. "Higher Education in the GCC: Linkages and Independence." *Gulf Research Meeting.* http://gulfresearchmeeting.net/index.php?pgid=Njk=&wid=MTA2&yr=2015.

Adely, Fida. 2004. "The Mixed Effects of Schooling for High School Girls in Jordan: The Case of Tel Yahya." *Comparative Education Review: Special Issue on Global Trends in Comparative Research on Gender and Education* 48 (4): 353-373. http://www.jstor.org.ezproxy.lib.purdue.edu/stable/10.1086/423361.

Agarwal, Anant. 2014. "Queen Rania Foundation Launches Edraak, a MOOC Portal for the Arab World." *edX News*, May 19. https://www.edx.org/blog/queen-rania-foundation-launches-edraak.

Al-Bustani, Hareth. 2015. "NYUAD's class of 2014: Where Are They Now?" *The National*, March 31. http://www.thenational.ae/uae/education/nyuads-class-of-2014-where-are-they-now.

Al-Mubaraki, Abdullah A. S. 2011. "National and Global Challenges to Higher Education in Saudi Arabia: Current Development and Future Strategies." In *Higher Education in the Asia-Pacific* (Vol 36, Higher Education Dynamics series), edited by Simon Marginson, Sarjit Kaur, and Erlenawati Sawir, 413–430. Springer Netherlands. http://link.springer.com/chapter/10.1007/978-94-007-1500-4_22.

Al-Munajjed, Mona. 2009. "Women's Education in Saudi Arabia: A Way Forward." A Work in Progress Paper. Booz&Co Ideation Center Advance Look.

ArabWIC. 2014. *Arab Women in Computing Website.* http://arabwic.org/.

Aswad, Noor Ghazal, Georgeta Vidican, and Diana Samulewicz. 2011. "Creating a Knowledge-based Economy in the United Arab Emirates: Realising the Unfulfilled Potential of Women in the Science, Technology and Engineering Fields." *European Journal of Engineering Education* 36 (6): 559–570. http://doi.org/10.1080/03043797.2011.62417.

Baki, Roula. 2004. "Gender-Segregated Education in Saudi Arabia: Its Impact on Social Norms and the Saudi Labor Market." *Education Policy Analysis Archives* 12 (28). http://eric.ed.gov/?id=EJ852310.

Bebbington, Diane. 2002. "Women in Science, Engineering and Technology: A Review of the Issues." *Higher Education Quarterly* 56 (4): 360–375. http://doi.org/10.1111/1468-2273.00225.

Beraud, Andre. 2003. "A European Research on Women and Engineering Education (2001-2002)." *European Journal of Engineering Education* 28 (4): 435–451. http://doi.org/10.1080/0304379032000101818.

Boashash, Boualem. 2013. "Principles of Excellence in Engineering Education Applied to GCC/MENA Engineering Colleges and Comparison with Australia and France: A Thesis with Analysis, Results, Observations." In *2013 7th IEEE GCC Conference and Exhibition (GCC)*, 547–552. IEEE. http://doi.org/10.1109/IEEEGCC.2013.6705838.

Bradley, Karen and Francisco O. Ramirez. 1996. "World Polity and Gender Parity: Women's Share of Higher Education, 1965–1985." In *Research in Sociology of Education and Socialization* 11: 63–91.

Breslow, Lori, David E. Pritchard, Jennifer DeBoer, Glenda S. Stump, Andrew D. Ho, and Daniel T. Seaton. 2013. "Studying Learning in the Worldwide Classroom: Research into edX's First MOOC." *Research & Practice in Assessment* 8: 13–25. http://mooc.pku.edu.cn/source/resource/01.pdf.

Brush, Stephen G. 1991. "Women in Science and Engineering." *American Scientist* 79 (5): 404–419. http://www.jstor.org/stable/29774475.

Butler, Declan. 2015. "Saudi University Backs Slow Road to Modernization." *Nature*, February 3. http://www.nature.com/news/saudi-university-backs-slow-road-to-modernization-1.16840.

Calvert, John R. and Abdullah S. Al-Shetaiwi. 2002. "Exploring the Mismatch Between Skills and Jobs for Women in Saudi Arabia in Technical and Vocational Areas: the Views of Saudi Arabian Private Sector Business Managers." *International Journal of Training and Development* 6 (2): 112–124. http://doi.org/10.1111/1468-2419.00153.

Carrington, Bruce, Peter Tymms, and Christine Merrell. 2008. "Role Models, School Improvement and the 'Gender Gap'— Do Men Bring Out the Best in Boys and Women the Best in Girls?" *British Educational Research Journal* 34 (3): 315–327. http://doi.org/10.1080/01411920701532202.

Cech, Erin, Brian Rubineau, Susan Silbey, and Caroll Seron. 2011. "Professional Role Confidence and Gendered Persistence in Engineering." *American Sociological Review* 76 (5): 641-66. http://doi.org/10.1177/0003122411420815.

Ceci, Stephen J. and Wendy M. Williams. 2010. *The Mathematics of Sex: How Biology and Society Conspire to Limit Talented Women and Girls.* NY: Oxford University Press.

Charles, Maria. 2011. "What Gender is Science?" *Contexts of the American Psychological Association* 10 (2): 22-28. http://doi.org/10.1177/1536504211408795.

Charles, Maria and Karen Bradley. 2002. "Equal but Separate? A Cross-national Study of Sex Segregation in Higher Education." *American Sociological Review* 67 (4): 573-99. http://doi.org/10.2307/3088946.

Charles, Maria and Karen Bradley. 2009. "Indulging Our Gendered Selves? Sex Segregation by Field of Study in 44 Countries." *American Journal of Sociology* 114 (4): 924-76. http://doi.org/10.1086/595942.

Data Centre. (n.d.). Retrieved January 18, 2015, from http://www.uis.unesco.org/datacentre/pages/default.aspx.

DeBoer, Jennifer. 2007. "Perspectives on Engineering Education in Tunisia after Fifty Years of Statehood: Opinions on Educational Quality and a Structural Analysis." *6th ASEE Global Colloquium.*

—. 2012a. "Twentieth-Century American Education Reform in the Global Context." *Peabody Journal of Education* 87 (4): 416–435. http://doi.org/10.1080/0161956X.2012.705136.

—. 2012b. "Why the Fireworks?: Theoretical Perspectives on the Explosion in International Assessments and Implications for the Gulf." In *Workshop 10 The Impact of Large-Scale Datasets on Evidence-Based Educational Policymaking and Reform in the Gulf States.*

—. 2016. "Perspectives on Engineering Education Quality in Tunisia After 50 Years of Statehood." In *Advances in Engineering Education in the Middle East and North Africa*, edited by Mahmoud Abdulwahed, Mazen O. Hasna, and Jeffrey E. Froyd. Cham: Springer International Publishing. http://link.springer.com/10.1007/978-3-319-15323-0.

DeBoer, Jennifer and Lori Breslow. 2014a. "National Differences in an International Classroom." http://odl.mit.edu/events/lori-breslow-jennifer-deboer-national-differences-in-an-international-classroom/.

—. 2014b. "Virtual Classrooms in a Real World Context." *Presented at the MOOCs 4D: Potential at the Bottom of the Pyramid.* Philadelphia, PA.

DeBoer, Jennifer and Glenda S. Stump. 2014. "National Differences in an International Classroom." In *Proceedings of the Fourth International Conference on Learning Analytics and Knowledge*, 281–282. New York, NY, USA: ACM. http://doi.org/10.1145/2567574.2567602.

Diekman, Amanda B., Elizabeth R. Brown, Amanda M. Johnston, and Emily K. Clark. 2010. "Seeking Congruity Between Goals and Roles: A New Look at Why Women Opt Out of Science, Technology, Engineering, and Mathematics Careers." *Psychological Science* 21 (8): 1051-57. http://doi.org/10.1177/0956797610377342.

Doumato, Eleanor Abdella. 2003. "Education in Saudi Arabia: Gender, Jobs, and the Price of Religion." In *Women and Globalization in the Arab Middle East: Gender, Economy, and Society*, edited by Eleanor Abdella Doumato and Marsha Pripstein Posusney, 243. Lynne Rienner Publishers. https://www.rienner.com/title/Women_and_Globalization_in_the_Arab_Middle_East_Gender_Economy_and_Society.

edX. 2014. "Saudi Arabia and edX Join Forces to Bridge the Gap Between Education and Employment in the Arab World." 2014. *edX Press*, July 15. https://www.edx.org/press/saudi-arabia-edx-join-forces-bridge-gap.

Faulkner, Wendy. 2000. "Dualisms, Hierarchies, and Gender in Engineering." *Social Studies of Science* 30 (5): 759-792. SAGE Publications. http://doi.org/10.1177/030631200030005005.

Felder, Dell and Mirka Vuollo. 2008. *Qatari Women in the Workforce.* (SSRN Scholarly Paper No. ID 1291424). Rochester, NY: Social Science Research Network. http://papers.ssrn.com/abstract=1291424.

Fouad, Nadya A. and Romila Singh. 2011. "Stemming the Tide: Why Women Leave Engineering." Center for the Study of the Workplace (report). University of Wisconsin-Milwaukee. http://www.studyofwork.com/files/2012/10/NSF_Report_2012-101d98c.pdf.

Gais, Hannah. 2014. "Saudi Arabia Gets MOOC'd Up | Al Jazeera America." *Al Jazeera America*, September 2. http://america.aljazeera.com/opinions/2014/9/saudi-arabia-massiveopenonlinecoursesgendersegregation.html.

"GCC Women: Challenging the Status Quo." 2012. Special Report. Al Masah Capital Management Limited. http://almasahcapital.com/uploads/report/pdf/report_19.pdf.

Gillard, Hazel, Debra Howcroft, Natalie Mitev, and Helen Richardson. 2008. "'Missing Women': Gender, ICTs, and the Shaping of the Global Economy." *Information Technology for Development* 14 (4): 262–279. http://doi.org/10.1002/itdj.20098.

Gordts, Eline. 2015. "Queen Rania: Have You Heard About the Arab World's 'Startup Spring?'" *The Huffington Post*. http://www.huffingtonpost.com/2015/03/06/jordan-rania-_n_6818214.html.

Hamdan, Amani. 2005. "Women and Education in Saudi Arabia: Challenges and Achievements." *International Education Journal*, 6 (1): 42-64. http://eric.ed.gov/?id=EJ854954.

Havergal, Chris. 2015. "Staying Away from Politics." *Inside Higher Ed*, March 12. https://www.insidehighered.com/news/2015/03/12/president-kaust-says-he-wont-criticize-saudi-policies.

Hill, Catherine, Christianne Corbett, and Andresse St Rose. *Why so few? Women in Science, Technology, Engineering, and Mathematics*. American Association of University Women. 1111 Sixteenth Street NW, Washington, DC 20036, 2010. http://eric.ed.gov/?id=ED509653.

Ho, Andrew Dean, Isaac Chuang, Justin Reich, Cody Austun Coleman, Jacob Whitehill, Curtis G. Northcutt, and Rebecca Petersen. 2015. *HarvardX and MITx: Two Years of Open Online Courses Fall 2012-Summer 2014* (SSRN Scholarly Paper No. ID 2586847). Rochester, NY: Social Science Research Network. http://papers.ssrn.com/abstract=2586847.

Hosea, Leana. 2014. "Why do Women Outnumber Men in Technology in the Gulf?" *BBC News*, February 8. http://www.bbc.com/news/business-26083215.

Hyde, Janet S., Sara M. Lindberg, Marcia C. Linn, Amy B. Ellis, and Caroline C. Williams. 2008. "Gender Similarities Characterize Math Performance." *Science* 321 (5888): 494–495. New York, N.Y. http://doi.org/10.1126/science.1160364.

International Labour Organization Organisation for Economic Co-operation and Development World Bank Group. 2014. *G20 Labour Markets: Outlook, Key Challenges and Policy Responses*. Report prepared for the G20 Labour and Employment Ministerial Meeting, Melbourne, Australia. http://www.oecd.org/g20/topics/employment-and-social-policy/G20-labour-markets-outlook-key-challenges-and-policy-responses.pdf.

ISCED: International Standard Classification of Education. (n.d.). http://www.uis.unesco.org/Education/Pages/international-standard-classification-of-education.aspx.

IGU/UNESCO. 2012. "Roundtable on Women in Engineering Concept Paper." http://members.igu.org/old/IGU%20Events/other-igu-events/unesco-igu-workshop-on-women-in-engineering/transcript-igu-unesco-women-in-engineering_stm.pdf.

Iredale, Robyn. 2005. "Gender, Immigration Policies and Accreditation: Valuing the Skills of Professional Women Migrants." *Geoforum* 36 (2): 155-166. http://doi.org/10.1016/j.geoforum.2004.04.002.

Islam, Samira Ibrahim. 2014. "Saudi Women: Opportunities and Challenges in Science and Technology." *Education Journal*, 3 (2): 71-78. http://doi.org/10.11648/j.edu.20140302.15.

Ismail, Maimunah and Mariani Ibrahim. 2008. "Barriers to Career Progression Faced by Women: Evidence from a Malaysian Multinational Oil Company." *Gender in Management: An International Journal*, 23 (1): 51 – 66. http://doi.org/10.1108/17542410810849123.

Jordan News Agency. 2015. "Queen Rania: The West Should Invest in Arab Creative Talents." *Petra, Jordan News Agency*. Amman. http://petra.gov.jo/Public_News/Nws_NewsDetails.aspx?lang=2&site_id=1&NewsID=185648&CatID=13.

Kam, Moshe. 2005. "Why Won't Jane Go to Engineering School (Hint: Jane is Not Dumb)." *The IEEE Institute*, 29 (4): 16-17. http://moshekam.org/commentary/files/Why%20Won't%20Jane%20Go%20to%20Engineering%20School.pdf.

Khreisat, Laila. 2009. "The Under-Representation of Women in Information Technology and Computing in the Middle East: A Perspective from Jordan." *Technology in* Society 31 (3): 287-294. http://doi.org/10.1016/j.techsoc.2009.06.006.

Lagesen, Vivian Anette. 2008. "A Cyberfeminist Utopia? Perceptions of Gender and Computer Science among Malaysian Women Computer Science Students and Faculty." *Science, Technology & Human Values* 33 (1): 5-27. http://doi.org/10.1177/0162243907306192.

Luke, Carmen. 2002. "Globalization and Women in Southeast Asian Higher Education Management." *Teachers College Record* 104 (3): 625-662.

MacKenzie, Jennifer. 2014. "The Real Transformers: How Women's Daring is Advancing Economies." *Barakabits.* http://www.barakabits.com/2014/03/real-transformers-womens-daring-advancing-economies.

Mafeo, Lauren. 2013. "Saudi Girls are More Likely to Study Science Than American Girls. Here's Why." *PolicyMic.com.* http://mic.com/articles/75819/saudi-girls-are-more-likely-to-study-science-than-american-girls-here-s-why.

Marchand, Gwen C. and Gita Taasoobshirazi. 2012. "Stereotype Threat and Women's Performance in Physics." *International Journal of Science Education* 35 (18): 3050-61. http://dx.doi.org/10.1080/09500693.2012.683461.

Matthews, Mark. 2013. "What's Her Secret?" *PRISM*, September. Washington DC: American Society for Engineering Education. http://www.asee-prism.org/featured-story-2/.

Mellström, Ulf. 2009. "The Intersection of Gender, Race and Cultural Boundaries, or Why is Computer Science in Malaysia Dominated by Women?" *Social Studies of Science* 39 (6): 885-907. http://doi.org/10.1177/0306312709334636.

Metcalfe, Beverly Dawn. 2008. "Women, Management and Globalization in the Middle East." *Journal of Business Ethics* 83 (1): 85–100. http://doi.org/10.1007/s10551-007-9654-3.

Morgan, Carolyn Stout. 1992. "College Students' Perception of Barriers to Women in Science and Engineering" *Youth and Society* 24 (2): 228-236. http://doi.org/10.1177/0044118X92024002006.

Morrison, Andrew R., Dhushyanth Raju, and Nistha Sinha. 2007. "Gender Equality, Poverty and Economic Growth." *The World Bank Gender & Development Group*. World Bank Publications.

Murphey, Emma C. 2003. "Women in Tunisia: Between State Feminism and Economic Reform." In *Women and Globalization in the Arab Middle East: Gender, Economy, and Society*, edited by Eleanor Abdella Doumato and Marsha Pripstein Posusney. Lynne Rienner Publishers.

Naser, Kamal, Wojoud Rashid Mohammed, and Rana Nuseibeh. 2009. "Factors that Affect Women Entrepreneurs: Evidence from an Emerging Economy." *International Journal of Organizational Analysis*, 17 (3): 225-247. http://doi.org/10.1108/19348830910974932.

National Science Foundation. 2011. "Women, Minorities, and Persons with Disabilities in Science and Engineering." NSF 11-309, Arlington, VA. http://www.nsf.gov/statistics/wmpd/digest/theme2_1.cfm.

OECD. 2012. International Migration Outlook 2012, Share of Women in Migration Flows, 2000-10. DOI:10.1787/migr_outlook-2012-graph18-en.

O'Connell, Dan. 2013. "Queen Rania Foundation Partners with edX to Create First MOOC Portal for the Arab World." *edX Press*, November 8. https://www.edx.org/press/queen-rania-foundation-partners-edx.

Onsman, Andrys. 2011. "It is Better to Light a Candle than to Ban the Darkness: Government Led Academic Development in Saudi Arabian Universities." *Higher Education*, 62 (4): 519–532. http://doi.org/10.1007/s10734-010-9402-y.

"Percentage of Women Earning Engineering Degrees: A Decade Long Flatline." 2015. A Data Report. Washington DC: American Society for Engineering Education.

PISA. 2012. Programme for International Student Assessment; OECD, PISA 2012 Database. http://pisa2012.acer.edu.au.

Quamar, Md Mudassir. 2013. "Education as a Ladder for Saudi Women: An Overview." *Journal of Arabian Studies* 3 (2): 265–277. http://doi.org/10.1080/21534764.2013.863680.

Ramirez, Francisco O. and Christine Min Wotipka. 2001. "Slowly but Surely? The Global Expansion of Women's Participation in Science and Engineering Fields of Study, 1972-92." *Sociology of Education* 74 (3): 231-251. http://doi.org/10.2307/2673276.

Regets, Mark C. 2007. "Research Issues in the International Migration of Highly Skilled Workers: A Perspective with Data from the United States." Working Paper, SRS 07-203, June 2007, Division of Science Resources Statistics, National Science Foundation.

Riegle-Crumb, Catherine, Barbara King, Eric Grodsky, and Chandra Muller. 2012. "The More Things Change, the More They Stay the Same? Prior Achievement Fails to Explain Gender Inequality in Entry into STEM College Majors over Time." *American Educational Research Journal* 49 (6): 1048-1073. http://doi.org/10.3102/0002831211435229.

Sagebiel, Felizitas and Jennifer Dahmen. 2006. "Masculinities in Organizational Cultures in Engineering Education in Europe: Results of the European Union Project WomEng." *European Journal of Engineering Education* 31 (1): 5–14. http://doi.org/10.1080/03043790500429922.

Searcey, Dionne. 2014. "A Conundrum for Saudis: Women at Work." *The New York Times*, November 28. http://www.nytimes.com/2014/11/29/business/a-conundrum-for-saudis-women-at-work.html.

Seymour, Elaine and Nancy M. Hewitt. 1997. *Talking About Leaving: Why Undergraduates Leave the Sciences*. Boulder, CO: Westview Press.

Shavit, Yossi, Richard Arum, and Adam Gamoran. 2007. *Stratification in Higher Education: A Comparative Study*. Stanford, Calif.: Stanford University Press. http://www.sup.org/books/title/?id=8735.

Shehadi, Ramez, Leila Hoteit, Kamal Tarazi, Abdulkader Lamaa. 2011. "Educated, Ambitious, Essential: Women Will Drive the GCC's Future." A Special Report. In *Perspective*. Booz & Company. http://www.strategyand.pwc.com/me/home/thought_leadership_strategy/40007869/50067873.

"Stemming the Tide: Why Women Leave Engineering" (2011, report).

Syed Zamberi Ahmad. 2011. "Evidence of the Characteristics of Women Entrepreneurs in the Kingdom of Saudi Arabia." *International Journal of Gender and Entrepreneurship* 3 (2): 123–143. http://doi.org/10.1108/17566261111140206.

Syeed, Nafeesa and Rahilla Zafar. 2014. *35 Entrepreneurs Making a Difference in the Arab World: Arab Women Rising*. Knowledge@Wharton, Wharton School of Business, University of Pennsylvania. http://kw.wharton.upenn.edu/arab-women-rising/.

Taibah, Maha K. 2015. "How Do You Say 'MOOC' in Arabic?" *edX* News, March 3. https://www.edx.org/blog/how-do-you-say-mooc-arabic.

Thom, Mary. 2001. *Balancing the Equation: Where Are Women and Girls in Science, Engineering and Technology?* National Council for Research on Women, 11 Hanover Square, 20th Floor, New York, NY 10005.

http://www.ncrw.org. http://eric.ed.gov/?id=ED455348.

Trends in International Mathematics and Science Study (TIMSS) - Overview. 2011. http://nces.ed.gov/timss/.

Trends in International Mathematics and Science Study (TIMSS) - Overview. 2014. http://nces.ed.gov/timss/.

Ulaby, Fawwaz. T. 2010. "KAUST: An International, Independent, Graduate Research University." In *University Research for Innovation*, edited by Luc Weber and James J. Duderstadt, 177-184. Economica.

UIS – UNESCO Institute for Statistics. 2008-2011. [www.uis.unesco.org]

UNESCO Institute for Statistics. 2008. Global Education Digest. Retrieved from [www.uis.unesco.org]

—. 2010. Global Education Digest. Retrieved from [www.uis.unesco.org]

—. 2011. Global Education Digest. Retrieved from [www.uis.unesco.org]

—. 2011. Country and Regional Profiles. Retrieved from [www.uis.unesco.org]

United Nations Educational, Scientific and Cultural Organization. 2012. World Atlas of Gender Equality in Education. Paris, France.

—. 2014. EFA Global Monitoring Report. Teaching and Learning: Achieving Quality for All.

Watson, Karan, Jorja Kimball, Karen Frair, and Mary Rowland-Anderson. 2006. "Efforts to Retain Women and Minorities in Engineering: A Perspective of the Foundation Coalition Project." In *Women in Engineering ProActive Network.*

Weber, Alan S. 2010. "Web-Based Learning in Qatar and the GCC States." Occasional Paper No. 5. Center for International and Regional Studies Georgetown University School of Foreign Service in Qatar. https://repository.library.georgetown.edu/handle/10822/558293.

Xie, Yu and Kimberlee A. Shauman. 2003. *Women in Science: Career Processes and Outcomes.* Cambridge, MA: Harvard University Press.

Zamberi Ahmad, Syed. 2011. "Evidence of the Characteristics of Women Entrepreneurs in the Kingdom of Saudi Arabia: An Empirical Investigation." *International Journal of Gender and Entrepreneurship* 3 (2), 123-143. http://doi.org/10.1108/17566261111140206.

Zeid, Amir and Rehab El-Bahey. 2011. "Impact of Introducing Single-Gender Classrooms in Higher Education on Student Achievement Levels: A Case Study in Software Engineering Courses in the GCC Region." In *Frontiers in Education Conference (FIE), 2011* (pp. T2H–1–T2H–6). http://doi.org/10.1109/FIE.2011.6142921.

Zghal, Riadh. 2006. "Les Femmes Ingénieurs en Tunisie: Représentations, Roles et Contribution." Tunis: CREDIF.

Zoli, Corri, Shobha Bhatia, Valerie Davidson, and Kelly Rusch. 2008. "Women: Engineering and Leadership." *Synthesis Lectures on Engineers, Technology, and Society* 3 (1): 1-106. San Rafael, CA: Morgan and Claypool Publishers. DOI: 10.2200/S00111ED1V01Y200804ETS005

"Masdar, in itself, needs to generate a new type of knowledge."

" [Masdar Institute] is really a milestone of what can be the future of the Emirati people."

"It all has to be kind of happening at the same time."

Three Masdar Institute faculty[1]

5

Co-Producing Emirati Science and Society at Masdar Institute of Science and Technology

Sebastian Pfotenhauer

Introduction: A Mandate for Science in the Emirates

In the eyes of the world, Abu Dhabi and the United Arab Emirates have come to be defined principally by their resource abundance, which have been both a boon and a bane for the region. The source of enormous wealth since the 1960s, oil and gas exports have propelled prosperity in Abu Dhabi to unprecedented heights – including one of the highest GDP-per-capita incomes and largest sovereign wealth funds of the world. On the other hand, the booming oil industry and the associated high pace of development have led to a range of well-documented economic, social, and environmental challenges, which have taken an increasingly firm hold on development discourses and questions of identity in the young nation (Government of Abu Dhabi 2008).

These challenges include first and foremost a monolithic economic dependency on the oil and gas market. While the structure of Abu Dhabi's economy has shifted in recent years towards other sectors, hydrocarbon exports are still responsible for 55% of Abu Dhabi's GDP of USD 260B (SCAD 2014), with much of the remaining 45% being fueled by downstream oil wealth. As a result, the Emirate continues to be extremely vulnerable to the boom-and-bust cycles of commodity markets, which in conjunction with a tenaciously low oil price in recent years and the looming exhaustibility of the hydrocarbon resources have resulted in increasingly urgent calls for diversification. Yet, structural changes to the economy are difficult: The challenges of a primarily resource-based economy have been discussed widely in the literature and include, for example, elements of the 'Dutch Disease' (i.e. currency appreciation due to commodity export-driven cash injections, with a subsequent decline in competitiveness in other domestic sectors), the crowding out of skilled labor and investment from other industries (particularly manufacturing), social and political reform lock-ins, governance challenges around rent-seeking tendencies, and other effects commonly subsumed under the term "resource curse" (Corden & Neary 1982; Neary & Wijnbergen 1986; Gelb 1988; Auty 1994; Karl 1997; Jensen & Wantchekon 2004; Morrison 2006; Dunning 2008; Ross 2013).

Second and related, its oil-based economy and geographic location make Abu Dhabi highly dependent on trends and developments in one technological sector: energy. As energy consumption around the globe continues to soar – particularly with the emergence of new economic heavyweights such as China and India – domestic energy consumption in prosperous Abu Dhabi has been soaring, too, creating challenges for the government to meet both local energy demand through cheap fossil fuel and long-term global supply commitments. At the same time, new energy sources are coming online, both in the form of renewables and "unconventional" oil sources such as shale oil and gas, putting a question mark behind the true economic value of Abu Dhabi's hydrocarbon resources. These developments have prompted the Abu Dhabi government to invest heavily in other energy sources, both renewables and nuclear power (WNA 2014), and strive to reduce domestic consumption (which hitherto has not been a national priority). Yet, renewable energy technology, nuclear power, and energy conservation measures require a fundamentally different technological basis than conventional oil extraction. Research has shown that a country's economic development prospects are strongly correlated with the diversity of its available technological bases, which has frequently prevented resource-based economies from advancing despite the abundance of financial capital (Hidalgo et al. 2007; Hausmann & Hidalgo 2011). This recasts Abu Dhabi's need for economic diversification as an imperative for technological diversification, which in turn mandates the creation of strong scientific basis.

Third, the reliance on extractive industries, the lavish use of fossil fuel for domestic energy consumption, and soaring living standards have led to a range of environmental and other sustainability challenges in Abu Dhabi. Low-efficiency housing, a car-centric transportation system, and an inefficient agriculture sector (most food is imported, and water desalinated, on a daily basis) have amounted to one of the highest carbon footprints per capita in the world (The World Bank 2014; AboulNaga & Elsheshtawy 2001; El Din et al. 1994; Alderman 2010). This has sparked a growing demand for scientific knowledge and technological solutions in the domain of sustainability and advanced energy, with as specific focus on regional needs.

Fourth, the high pace of development has come with an excessive dependence on foreign labor across the entire skills spectrum, from low-wage South-Asian construction workers to Western-trained high-tech engineers and asset managers. In 2013, only 20% of the Abu Dhabi's population and 9% of the labor force (132,000 out of 1,443,700) were Emiratis (SCAD 2014). Technological expertise in particular – and hence the building blocks of a diversified knowledge economy – remains firmly in the hands of foreign companies and contractors, which poses question marks behind the country's diversification plans and its capability to address its environmental challenges with a domestic knowledge stock.

In 2006, the government of Abu Dhabi launched *Masdar,* a comprehensive 'flagship' initiative to address this fourfold set of intertwined challenges: economic diversification, technological diversification, environmental sustainability, and a skilled domestic labor force. The spearhead of a broader "core commitment" on the part of the government "to build a sustainable and diversified, high value-added economy by 2030" (Government of Abu Dhabi 2008), Masdar is envisioned as "a catalyst for the economic diversification of the emirate, [..] driving new sources of income for the emirate and strengthening Abu Dhabi's knowledge-based economic sectors" (Masdar Corporate 2014b).

The heart of this initiative is *Masdar Institute of Science and Technology* (or *Masdar Institute* for short), the first graduate-level research university of the country. Founded in 2006, Masdar Institute has been tasked with leading the country's "evolution as a leader in global energy from a provider of fossil fuels to a developer of alternative energy and clean technologies" to make "Abu Dhabi the preeminent source of renewable energy knowledge, development and implementation, as well as the world's benchmark for sustainable development" (MIST 2011). Masdar Institute is envisaged as "a world-class research and education environment that attracts and nurtures future leaders and critical thinkers in science and technology with particular focus on advanced energy and sustainability" (ibid.). It is to "develop [the] country's most crucial resources: its people [..] to feed the UAE and the region with highly talented individuals and research expertise, enhancing economic development and diversification for the benefit of future generations." Deriving from the Arabic word for "source," Masdar Institute is seen at once as the spring of a future with smaller dependence on oil exports, a flourishing clean-tech economy, greater environmental responsibility, and a highly educated populace capable of sustaining a modern knowledge-based economy.

In this chapter, I will explore this proposed role of Masdar Institute in the socioeconomic transformation of the Abu Dhabi and the GCC region. I will argue that the Institute's inception represents a key moment for the Emirates in which a *scientific enterprise* and a *new social order* are being envisioned and forged in tandem, and widely shaped through their mutual interaction. What goes on at Masdar Institute (and, by extension, similar flagship initiatives in the GCC region) embodies a range of social and political values, which are in turn re-inscribed into society through the envisaged unfolding of science. This interaction between science and society in the making of a novel institutional landscape warrants close attention if one wants to understand and anticipate the specific development trajectory on which a country is embarking.

Building Science, Building Society: Co-Production as a Framework for Analyzing Research Universities in Late-Developing Countries

How should one study the mutually constitutive relationship between science and society? From the dawn of Enlightenment, science and technology have shaped the way we see, live in, and transform the world, both individually and collectively. Conversely, history has witnessed how different societies give rise to different scientific cultures, with distinct patterns of *what* and *how* they research, and how they choose to deploy the fruits of technological progress. The ever-growing interfusion of social life with science and technology is widely considered a trademark of modernity (Beck 1992; Giddens 1991; Latour 1993; Jasanoff 2004b), which has brought about tremendous improvements in prosperity, living conditions, and human control over nature. At the same time, science and technology have given rise to new forms of inequality, risks, concentration of power, and human vulnerability. They have become essential tools of governance and state-making, both through their implied organizational rationality as well as their inseparability from the articulation and realization of national socio-political visions (Foucault 1979; Ezrahi 1990; Scott 1999; Jasanoff & Kim forthcoming). Today, the endless advancement of science and technology is seen by many as a matter of economic necessity (Schumpeter 1934; Bush 1945; Solow 1956; Landau & Rosenberg 1986; Romer 1994).

Institutions of knowledge creation – and universities in particular – have played a central role in this manifoldly intertwined history of science and society. From the first modern universities with their focus on natural philosophy and divine knowledge (Rudy 1984; Rüegg & Ridder-Symoens 2003); to the Humboldtian research university with its two-fold mission of teaching and research, enmeshed in the wants and needs of an emancipating citizenry (Humboldt 2002; Fichte 1971a); to the American land grant colleges with their economic and democratic overtones (Geiger 1986; Jewett 2014) and the 20th century shifts from elite to mass higher education, from open-ended to mission-oriented research, and from the lab bench to big science (Trow 2002; Stokes 1997; Galison & Hevly 1992; Geiger 1993); all the way to the turn of the century with its focus on innovation, situated knowledges and embedded science (Gibbons et al. 1994; Etzkowitz & Leydesdorff 2000; Conceição & Heitor 1999; Thorp & Goldstein 2010) – universities have always been both a vehicle for, and representation of, their respective societies.

Scholars from the field of Science and Technology Studies (STS) have theorized the mutually generative relationship between science and society in a variety of ways, and with it the key role of knowledge-producing institutions therein. One framework that has recently gained prominence is the concept of *co-production*, introduced by Jasanoff (2004a) as "a language for hybrids" that recognizes "that the production of order in nature and society has to be discussed in an idiom that does not, even accidentally and without intent, give primacy to either." Co-productionist analysis guides our view to the ways in which scientific knowledge and technological artifacts create and modify social order; conversely, it helps us unravel how the production of science and technology itself, and with it the very institutions responsible for their creation, are shaped by social and political factors. "Scientific knowledge," Jasanoff writes, "both embeds and is embedded in social practices, identities, norms, conventions, discourses, instruments, and institutions – in short all the building blocks of what we term the *social*" (Jasanoff 2004b).[2]

So-called 'late-developing' countries (including those in the GCC region) with their nascent science landscapes and first-of-their-kind research universities are particularly tempting for co-productionist analysis. Here, sweeping governmental mandates for techno-economic development frequently collide with a set of social, political, and environmental conditions profoundly different than those encountered in the highly technologized societies of the West. As we shall see, the sheer number of 'firsts' encountered in the creation of Masdar Institute make the co-productionist nature of institution-building for science and technology all the more pronounced. What is more, technoscientific initiatives in these countries frequently combine a de-facto dependency on foreign technological capabilities with a rhetoric of scientific emancipation and continued struggles for a seat at the global scientific table.[3] In studying Masdar, this chapter thus follows in the footsteps of co-productionist research on changing institutional landscapes of science and technology, such as Krige's account on American hegemony in the post-war reconstruction of European science (Krige 2008) or Dennis' analysis of Vannevar Bush's rhetorically powerful yet institutionally contested vision for post-war science (Dennis 2004). Yet, in contrast, it shifts the locus of analysis to a 'late-developing' country demonstrating that nascent science landscapes (and particularly start-up universities with national flagship character) represent profoundly generative subjects for co-productionist inquiry.[4] Looking in particular at the intersection of two 'sites' of co-production suggested by Jasanoff (2004a) – the making of *institutions* and the making of *identity* – I will show that the emerging science-society model in Masdar is markedly different from that of the liberal democracies of the West.

Inventing Emirati Science: Sustainability and Local Relevance Under a Corporate Umbrella

To explore how science as constructed at Masdar Institute embodies distinctly Emirati characteristics, it stands reason to begin by assessing the Institute's broader institutional context. Masdar Institute is but one of four strands of the Masdar initiative, designed to serve an interlocking set of development functions (Reiche 2010): *Masdar City* is an urban development project and "model city" that is envisioned as a place fully reliant on renewable energy sources, car-free, and with minimal carbon and waste footprint. Surrounding

Masdar Institute, the city is projected to accommodate some 40,000 residents and 50,000 more commuting workers (primarily high-skilled scientists, engineers, and entrepreneurs) as well as 1500 firms specialized on energy and sustainability-related manufacturing and investment (Hopwood 2010; Nader 2009). Masdar City is expected to become a 'living laboratory' for advanced sustainability technologies emanating from Masdar Institute and elsewhere in the world.[5]

Masdar Clean Energy, the third strand of the initiative, is a corporate developer of medium to large-scale sustainable energy projects in the region and intended as an important customer of the technologies emerging from Masdar Institute and City. *Masdar Capital*, finally, is the venture capital arm of Masdar with a mission to pursue high-risk technology commercialization in the area of clean technologies. The developments and investments made by these two entities have attracted considerable attention: In Abu Dhabi, Masdar co-funded and co-developed *Shams 1*, a 100MW concentrated solar power project and joint venture with Abengoa Solar and Total, which at the time of its opening was the world's largest solar project. Shams 1 was also the world's first concentrated solar power plant to be registered under the UN Clean Development Mechanism as a project eligible for carbon credits. In an another demonstration project, Masdar and ADNOC (the Abu Dhabi National Oil Company) launched a joint venture to develop a commercial-scale carbon capture, usage and storage facility, which includes a carbon dioxide compression unit at the Emirates Steel factory, a 50km pipeline, and injection sites for enhanced oil recovery in the ADNOC oil fields, finally storing the CO2 underground.

In its four-pronged structure and commercial orientation, Masdar could be characterized as a massive research-to-demonstration-to-valorization-to-habitation pipeline around an advanced energy and sustainability theme. That is, a sectorally specific innovation ecosystem that is to develop and implement commercial solutions to solve its host society's most pressing challenges. Masdar is at once the government's bet on, and touchstone for, a different economic, environmental, technological, and social future for the entire region.

Putting Masdar Institute at the center of this commercial ecosystem follows the common policy rationale that universities can act as key drivers of innovation and regional economic transformation, seen as capable to simultaneously address the entire "knowledge triangle" of education, research, and innovation (Mansfield 1991; Mansfield & Lee 1996; Varge 1998; Conceição & Heitor 1999; Etzkowitz 2002; Mowery & Sampat 2006; Youtie & Shapira 2008; Thorp & Goldstein 2010; Mercier-Laurent 2011; Butler & Gibson 2011; Technopolis 2012; EC 2012). Yet, this institutional embedding has direct implications for the scientific enterprise developed at Masdar Institute. First, while Masdar Institute itself is officially a not-for-profit organization, it is "still part of a profit-making enterprise" – not a science effort *per se* – and hence will ultimately be "evaluated in economic terms," a Masdar faculty suggests. Masdar itself is a subsidiary of the state-owned Mubadala Investment and Development Company, and as such an integral part of an economic development project (Fig. 1). For some Masdar faculty, this economic superstructure represents "a challenge, because you have to prove your work to a profit-making enterprise. And it's not quite clear to [the government], as you can understand, something that has a return horizon of 20 years, versus somebody who's trying to make a buck three years down the line. [At times] I feel like I'm writing a business plan more than a scientific proposal."

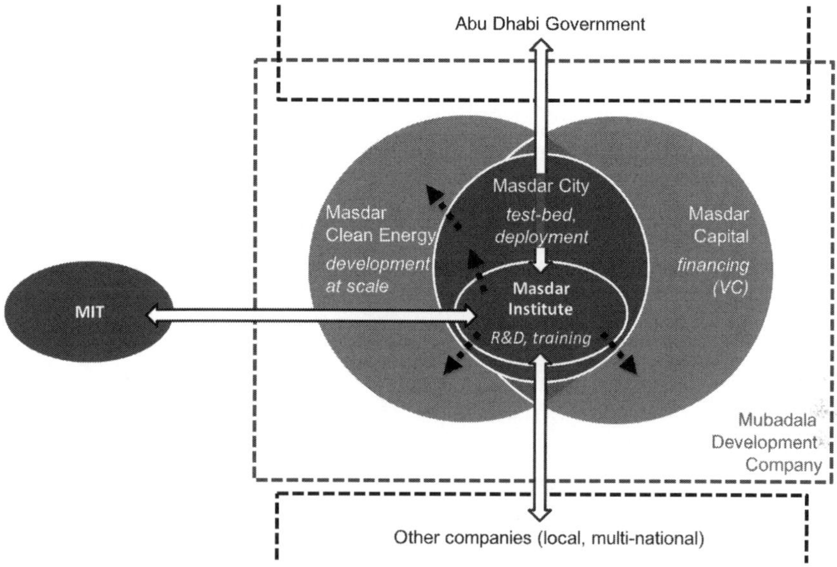

Figure 1: Masdar ecosystem, a four-pronged innovation ecosystem that includes Masdar Institute, Masdar City, Masdar Clean Energy, and Masdar Capital.

Yet, Masdar Institute's economic agenda is neither hidden nor a gradual recalibration of a formerly 'purely' scientific institution, as has been observed and problematized repeatedly for Western universities (Croissant & Restivo 2001; Gibbons et al. 1994; Etzkowitz & Leydesdorff 2000). On the contrary, Masdar openly embraces its corporate umbrella and its ties to national development efforts. One Masdar faculty explains that already prior to his hire, "I kind of viewed Masdar as a semi-industrial university, like it's somewhere between an R&D department at Intel and a full-fledged university." Moreover, faculty are not merely at the receiving end of this economic agenda, but may shape Abu Dhabi's broader development through their research: "We've never been directed, do this, this, this, and this, but it's always 'prove that you're relevant!' I don't really think the government [..] has a clear idea of where they want the research directions to go. So we've been in the position of kind of leading that."

Second, in its strong sectoral orientation and focus on regional challenges, Masdar Institute is a deeply "mission-oriented" research organization (Stokes 1997). Research at Masdar is centered on three broad thrusts, all related to advanced energy and sustainability: Water, Environment and Health; Future Energy Systems; and Microsystems and Advanced

Materials.[6] What is more, these research domains all come with a strong local focus. As a senior administrator at Masdar puts it:

"It had to be activities that are of high impact and high relevance to Abu Dhabi and in line with our mission, [..] advanced energy and sustainability. So on the water side, water desalination is a major sector here, and a major need: [..] advanced membranes for water desalination [..] water resources, water use, and impacts of climate variability on water, Gulf water quality, various environmental indicators. So there's that on the water side, that's very easy to argue for relevance. It's just, you know, it's an incredibly water stressed area, and [..] we're dependent upon a source of energy to provide us with water. We also have, in the energy systems sort of theme, different solar technologies, PV and concentrated solar power, and that was kind of already the birth of the Masdar initiative. [..] But there are some interesting challenges that'll come out with solar in terms of managing the dust in the atmosphere, the humidity, and how the technology may need to be adapted to be able to handle that. One thing that we've pushed [..] as a research direction on the energy side is in the bio-mass and bio-fuels. But it's unique here, in that it's looking at biofuel [..] that you can irrigate with seawater. And so it's all seawater based plant material, whether it's algae or terrestrial plants as well, and then waste."

Masdar Institute's focus on local applicability is a critical component of Abu Dhabi's envisioned socioeconomic transformation. While sustainability and clean-tech research are globally on the rise, these technologies do frequently not apply to the Emirate's geographical or social sociotechnical conditions. For example, photovoltaic installations need to cope with extreme temperatures, persistent soiling from desert sands, and limited reliance on human labor during the day because of the heat. As one policy analyst notes: "Masdar is not only a partner for other research institutions, it's also a good partner even for suppliers to see how the equipment performs in this type of very arid climates and very high temperatures [..] and how to manage it without using water, which is extremely scarce."

The theme of local relevance and regional environmental conditions furthermore permeates all sponsored research projects at Masdar Institute, be they with governmental agencies (such as the Environmental Agency, the Dubai Supreme Council of Energy, or International Renewable Energy Agency IRENA), national firms and developers (such as the Abu Dhabi National Oil Company ADNOC, Etihad, and Mubadala), or multinational companies (including UOP-Honeywell, Boeing, Safran, and Siemens). For example, the Masdar Sustainable Bioenergy Research Consortium, jointly operated with Boeing, Etihad and UOP Honeywell, features a program on sustainable aviation biofuels derived from saltwater plants, and the development of an Integrated Seawater Energy and Agriculture System that combines biofuel feedstock cultivation with aquaculture production and mangrove silviculture. Another example, the Masdar Research Center for Renewable Energy Mapping and Assessment is developing national-level solar and wind resource mapping tools, and work towards a publicly accessible atlas of solar and wind resources. The Center was launched in collaboration with the International Renewable

Energy Agency (IRENA), the UAE Directorate of Energy and Climate Change, the Dubai Supreme Council of Energy and Environment Agency, the UAE National Center of Meteorology and Seismology (NCMS), Total (France), and ParisTech (France). In this sense, Masdar Institute is also serving a gateway for, and repository of, expert knowledge on local technology use in the GCC region.

Third, Masdar Institute's mission orientation is visible throughout its education programs, which include 9 Master's programs and an interdisciplinary PhD. While all education tracks are 'classical' engineering degrees (e.g. Chemical Engineering, Computer and Information Science, Mechanical Engineering), all are customized around energy and sustainability themes. For example, the Chemical Engineering program states as its learning objectives "an ability to identify and address current and future chemical engineering problems related to energy sources, generation, conversion and green chemical production within a broader framework of sustainable development," and "an ability to apply a multi-disciplinary approach to conceive, plan, design, and implement solutions to chemical engineering problems in the field of energy and sustainability" (MIST 2011). One faculty summarizes the mission orientation in education as follows: "Are these Emirati students trained on topics relevant to Abu Dhabi? A lot of them are."

Science at Masdar Institute, the country's first research university, is thus indeed highly specific to the national needs and institutional context of Abu Dhabi. Shaped in its research and education activities by a mandate for a *specific* technology focus and high regional relevance, Masdar Institute is not – and does not aim to be – a 'complete' research university. Its scientific portfolio is thoroughly geared towards one sector with a notable orientation towards application and development for customers such as *Masdar City, Masdar Clean Energy,* and various other corporate and governmental organizations. Consequently, Emirati scientists-to-be do not receive just any type of research-based graduate education at this national flagship institution, but a custom-tailored engineering education centered on energy and sustainability.

The observed confluence of academic, economic, and political, interests in the Masdar model seems to go even further than what is generally known as the "triple helix" model of innovation, where innovation is envisaged to occur in the intersection of university, industry, and government activitaties (Etzkowitz 2002; Etzkowitz 2008). This entanglement is perhaps best exemplified by the enrolment of Masdar Institute in the MIT Energy Initiative (MITEI), which is part of the broader Masdar-MIT partnership discussed below. MITEI is MIT's main cross-departmental unit for energy research (both conventional and sustainable), and its main interface with industrial R&D interests on energy matters. In an unusual move and after extended negotiations with MIT, Mubadala enlisted Masdar Institute – after all, an academic institution – as a *corporate* member and co-sponsor of MITEI, where it stands out among MITEI's other clients that mainly consist of multi-national firms such as Shell, BP, or Saudi Aramco. Yet, this role is consistent with the corporate understanding of Masdar and Mubadala, which officially call Masdar Institute its "research arm" (Masdar Corporate 2014a). One could argue that the government of Abu Dhabi literally funds and uses an academic institution (Masdar Institute) as the research arm of a state-owned company (Mubadala) to drive and diversify the economy and advance technological development for urban and energy infrastructure projects. As one senior administrator at MIT involved in Masdar put it: "In a sense, they [Masdar Institute] are an appropriate organ for the government to use as the equivalent of a corporate research arm here. [..] Some of the energy exploration and operating companies that are owned by the government [had] some level of technical expertise. But it's not high level, and it's not well centralized or coordinated. By far the best coordinated is the Masdar Institute."

Arguably, Masdar thus condenses the "innovation triple helix" into a single *triple point* of indistinguishability between government, industry, and academia. Masdar Institute is an integral part of a broader *corporatist-etatist agenda,* in which the government of Abu Dhabi applies corporate-managerial rationales and instruments – including state-owned industries and development corporations – to pursue strategic development objectives around economic and environmental sustainability. It speaks to a situation where science and society are not viewed as separate entities, but as inextricable and complementary parts of one and the same economic development agenda. The classical image of an autonomous "Republic of Science" (Polanyi 1962) where science takes place in an independent domain of inquiry unfettered by political, economic, or personal interests, does most certainly not apply in Abu Dhabi.

(Re-)Inventing Emirati Society: Identity and 'Caravans of Knowledge Workers' in Abu Dhabi's Transitory Model of Science-Society Relationships

So how does Masdar Institute's scientific, technological, and educational agenda fit with Abu Dhabi's society at large? To begin with, it has been observed that the structure of Abu Dhabi's economy, labor force, and social stratification poses constraints in terms of economic development options: Abu Dhabi has been called a capital-rich, labor-poor[7] state with extremely high per-capita income. 'Classical' development scenarios such as the build-up of a low-wage manufacturing sector that gradually works its way up to the innovation frontier (like Singapore, China, or other oil nations like Malaysia and Indonesia) are not feasible for the Emirate, both for the small size of the domestic labor force and the de-facto reduction in per-capita income that shift into a low-wage sector would imply. To maintain its current per-capita wealth, Abu Dhabi requires an economic leap that directly moves into low-labor, capital-intensive, high value-added industries capable of sustaining high wages. In other words, it requires an economic structure based on high science intensity and high educational attainment, one typically associated with research-based graduate degrees, for which the government hitherto had no vehicle. Masdar Institute was meant to change that.

Masdar Institute thus entails a vision for a new, more active role for Emirati citizens in securing the future of their country. It envisions an educated populace that is capable and willing to move from oil-subsidized personal wealth to science-based economic productivity and global competition – a populace that subscribes to the proclaimed imperative of a socioeconomic and environmental transformation, and that strives to reduce the country's dependency on foreign high-skilled labor. Above all, it expects a citizenry that sees science (and particularly sustainability science) as a key part of its future, both collectively and individually. This responsibility of a scientifically literate would-be-citizenry and their key role in the envisaged socioeconomic transformation rings through loud and clear in the Masdar Institute vision: "People are the true driving force behind Abu Dhabi's ambitions and Masdar Institute is strategically positioned to feed the UAE and the region with highly talented individuals and research expertise, enhancing economic development and diversification for the benefit of future generations to come. [..] [It will] shape today's ambitious youth into tomorrow's scientific pioneers" (MIST 2011).

6 years into its operation, Masdar Institute has indeed made significant strides in attracting local students: Of 189 new students enrolled in 2014, 47% were Emiratis, with many more stemming from other Arab-speaking countries. Regional traction continues to grow, with applications by UAE nationals rising from 85 in 2009 to 285 in 2014. What is more, despite the well-known difficulties to attract female students into engineering, females currently represent 44% of the student population at Masdar Institute, of which in turn 44% are UAE nationals. According to some Masdar faculty, this high turnout is owed to the fact that Emirati women are highly educated and keen on obtaining graduate education, but primarily interested in opportunities that permit them to stay in the country and close to their family.[8]

Yet, given the outsize ambitions of the Masdar initiative, it remains an open question to what extent Masdar Institute can single-handedly drive the envisioned socio-economic transformation. Even with enrolments soaring, it is unlikely that the projected tens of thousands of jobs in Masdar City can be created and filled in real-time by Masdar Institute alone. Moreover, it is unclear to what extent science jobs – albeit foreseeably well remunerated in the Emirate – will actually offer an appealing alternative to other high-paying jobs guaranteed to Emiratis by virtue of their citizenship. The sustainability of the current monarchy-based political system arguably depends on the government's continued ability to balance a democratic deficit by providing high-paying jobs and sufficient individual wealth among its citizens, the risky flipside of which has been witnessed in the uprisings of the Arab spring. The continuous need to inject wealth into the population, and the factual ability to do so in a country that exports of the order of 300 USD in oil and gas products per citizen per day, may thus work against the vision of an Emirati science class.[9] From this perspective, the Masdar initiative in general and Masdar Institute in particular could potentially reinforce, rather than reduce, Abu Dhabi's foreign labor force dependency at the level of knowledge-workers.[10]

One conceivable option to address the current shortage in domestic scientists and skilled labor could be immigration. Many countries around the world are offering preferred immigration status to highly skilled individuals and are recruiting foreign graduates from domestic universities into the labor force – most notoriously in the U.S. (Freeman 2014; Freeman 2005; Bound et al. 2009; Hunt & Gauthier-Loiselle 2008). Yet, as of yet, there exists no formal path to citizenship in the UAE, and the government is firmly upholding the differences between citizens and foreigners. Partly as a result, there exists no tenure for faculty at Masdar Institute. While Masdar Institute faculty are granted visas for an exceptional four years (as opposed to the annual work visa for the majority of migrant workers), expats can – in theory – be deported anytime for any reason.

This lack of a long-term perspective for Masdar Institute faculty and other foreign researchers could, on the one hand, bode problems for the prospects and sustainability of Abu Dhabi's nascent scientific landscape. For one, it potentially limits the degree of identification that faculty and students can develop with their institution and the country at large as most foreigners expect to leave Abu Dhabi at some point, and may thus contribute to a somewhat transitory mindset. For another, it may affect the nature of research undertaken at Masdar Institute, i.e. the degree to which researchers are willing to embark on truly long-term research projects. Together, this puts at risk the very knowledge base that Abu Dhabi's is trying to develop through Masdar Institute. A faculty explains:

"Because there really isn't a citizenship entry path, and there isn't also a sense of like a green card, permanent visa, as there is in the United States or in other western European nations, people don't necessarily feel like they're always going to stay. So, you know [..] they view it as an opportunity to maybe get some money or gain some experience, or whatever. And then they move on with their work. [..] I highly respect Emirati society and Emirati culture, and what they've built here. But it really very much seems like the people move in and out like the wind. [..] And certainly Emiratis stay, but 80 percent or 85 percent of the nation is not really staying. [..] I think this has all sorts of implications in terms of like quality of service of pretty much every government sector, public sector, private sector activity, and it also means a lot in terms of the ecosystem in which Masdar is going to now participate. [..] I mean, it's as simple as [..] procurement practices or subcontractors, or all of the supporting activities to make a world class research university function are either non-existent, or they are, you know, being put in place. And if they are put in place, then maybe the people who are running them [..], at best they have a great deal of experience abroad, but only have a couple of years of experience that is specifically for Abu Dhabi. Or at worst, maybe they don't have the experience at all, and they only have those two or three years of effectively being an assignment here, right? So, and that's at pretty much every level, from these construction workers at the bottom, all the way up to doctors."

On the other hand, estimating the prospects of Masdar Institute by way of comparison with Western institutional traditions might be misleading. In fact, it risks missing essential aspects of local social embedding on which emerging institutional configurations of knowledge production have always depended, all the more if they are intended to address local problems. As argued above, the institutional models of the West – from research universities in early 19th century Germany to American land grant institutions to today's entrepreneurial universities – were all co-produced as part of the particular science-society relationships of their time and place.

An alternative, arguably more realistic account of Masdar's projected continuous reliance on transitory knowledge-workers would have to account for its distinct Emirati identity: Here, the idea that leagues of traders – traders of knowledge, in this case – might pass through the Emirate, remain in the country at the invitation of the Sheikhs, and continue onwards after some time might appeal to a traditionally Bedouin society. One Masdar faculty suggests that "nations develop along what has historically been their point of success. So you know, the Emirati society is traditionally a Bedouin society, and they succeeded in terms of caravans, in terms of trade, and then now, I look at the success of Etihad Airways, of Emirates Airlines, of Dubai Port, and the ability to basically move large amounts of people and products in and out, it continues to be the success of the mission now, right?"

What is more, this bet on a continuous influx of knowledge and people relates to a deeper-rooted set of social and institutional premises in the UAE modernization model. This model assumes that many of the building blocks of a modern society can be 'purchased' and imported in the style of expensive commodities – to the metaphorical and literal middle of the desert, if need be – often with little regard for the commensurability between the social context from which these practices originate, and that to which they are being transplanted. In fact, Abu Dhabi has built quite a reputation for institutional imports, which include among others college education, fine arts, architecture, sports, and international organizations.[11] Masdar Institute, jointly established with MIT (cf. below), fits well within the trajectory. Indeed, it would be curious if institutions of science and innovation – two other trademarks of Western success – would be exceptions to this rule. A Masdar faculty suggests:

"Well, the best way to transform an economy, what you do is just buy a company, build a factory. What do you do [..] if it is a knowledge based economy that you'd like to build? Then you need to build a nucleus for generating knowledge which is innovative. [..] This nucleus of this Masdar Initiative is the Masdar Institute, the university, [..] and they went to the best in the business, which is MIT, to build a technology-based university that would be the nucleus of this transformation."

Yet, this language of "import" and "buying" should not be mistaken for a naïve account that Abu Dhabi is indeed trying to build "an MIT." To the contrary, the above analysis suggests that Masdar Institute is fundamentally different from MIT, shaped by a distinct vision for Abu Dhabi that determines what kind of science is needed and why, and how the new institution needs to reflect to the social, cultural, and economic idiosyncrasies of the emirate. This vision is most certainly not congruous with MIT's own understandings of what a research university is. A thorough analysis of how local identity and self-diagnosed needs enter into the co-construction of this relationship goes beyond the scope of this chapter (Pfotenhauer & Jasanoff 2015; Pfotenhauer 2015; Pfotenhauer et al. 2016).

It is perhaps not surprising, then, that Masdar is extremely *demonstrative* about its Emirati identity, both in the sense of *being representative of* the UAE vision and *showcasing* that the country is making headway towards this vision. For example, in 2013, the newly elected cabinet was jointly sworn in as part of the opening celebration of the Shams 1 solar power plant, co-developed by Masdar (Salem 2013), demonstrating publicly that the Abu Dhabi government stands united behind Masdar.[12] Identity is also expressed in the architecture of Masdar Institute itself, which is constructed in ornamental Arabic style and features contemporary re-interpretation of traditional Arabic architecture, such as a wind tower to channel cooling breezes to the ground level. In terms of public visibility, faculty are regularly spotlighted in the media, as are students and their achievements.[13] Moreover, Masdar Institute has become a globally highly visible flagship sustainability initiative and somewhat of a must-go site for government leaders, renowned scientists, and other celebrities interested in flagship sustainability efforts[14] – in part through a very active outreach and publicity strategy. In short, Masdar is caught up in an ongoing display and performance of Emirati identity and progress, and has become the standard-bearer of the governmental vision for the emirate's future. A faculty describes the national flagship character as follows: "Our nation, I mean, people here 10 years ago [..] had never thought [that they can have] an institute that will focus totally on research. [Masdar Institute] is really a milestone of, you know, what can be the future of the Emirati people. [..] You know, we can test, they can transform the region into a place where they're able to breed smart and passionate scientists."[15]

Masdar Institute thus engenders a different social model of science from that of the West: one that is less reliant on permanent institutional allegiance or long-term guarantees, as embodied by the tenure or public servant model of the West; one that needs a different kind of incentive structure for foreigners as well as locals, the latter of which are offered abundant social guarantees by other means; one that balances 'imports' against local identity and the co-construction of a social embedding; and one that presupposes an Emirati citizenry dedicated to refashioning the foundations of their country based on science and technology. Masdar embraces a more dynamic, trade-oriented model of knowledge production and exchange between a locals and 'caravans' of knowledge workers that may nonetheless be powerful and sustainable in its own terms.

This re-invention of Abu Dhabi's society does not come without trade-offs, however. The way in which a new society is being envisioned and fashioned around Masdar's agenda is reminiscent of Scott's (1999) analysis of top-down governmental schemes of social engineering around development projects and a technocratic production logic. As Scott demonstrates, these schemes have time and again run into challenges regarding adequate contextualization in preexisting social fabrics, and frequently resistance by the very public on whose behalf they are supposedly acting. Scott argues that the shoring up of sufficient financial, political, and ideological muscle behind developmental megaprojects typically comes at the expense of democratic participation and public appropriation, which in turn might undermine that these projects take organic roots. Ponzini's (2011) observation that the UAE future is being written in a top-down manner by an extremely small group close to the royal family whose members oftentimes straddle public and corporate offices would support such a reading.

Stabilizing Co-Production: Institution-Building in the Masdar-MIT Partnership

In light of the above, the creation of Abu Dhabi's nascent science landscape and associated social context presents itself as a highly tangled, if not fragile endeavor. On the one hand, Abu Dhabi's vision for economic, environmental, and social development is predicated on a flourishing scientific enterprise. Yet, how does create such an enterprise without having in place an socio-economic and institutional environment that could nurture it to begin with? How does one jumpstart and justify a locally rooted enterprise without having the local human resources? On the other hand, the success of Masdar's technoscientific ambitions presupposes interest and buy-in both by the country's citizens and the global community. Yet, how does one signal credibility for one's sociotechnical vision, and credible progress towards it, to local that does not yet embody and live this vision? How does one attract the desired knowledge workers and students to an institution that does not yet exist, and a place that has no track record? How do you let them research or educate them without having the necessary infrastructure in place?

Arguably, these concerns are shared by many institution-building efforts around science and technology, particularly in 'late-developing' countries. However, from a co-productionist perspective many of institution-building challenges pose a kind of chicken-and-egg problem in that they require a simultaneous change in both science *and* society. Indeed, many Masdar participants emphasize the regular experience of situations where challenges cannot be resolved on one side of the equation alone, where the way forward on one end is frequently unclear or blocked by the other end, and where the boundary between technoscientific and social activities become blurry. A faculty member illustrates: "In terms of readiness of jobs here, [..] it's kind of a cart and horse, where we're trying to create jobs at the same time we're trying to create the research and the graduates." Another faculty sums it up: "It all has to be kind of happening at the same time."

It is precisely from this perspective of co-productionist 'lock-in' that I wish to discuss the role of MIT-Masdar partnership in the creation of Masdar Institute. Above all else, MIT-Masdar is an act of facilitation and stabilization of co-productionist dynamics whereby Masdar Institute draws upon external resources and selective forms of expertise provided by MIT's to simultaneously advance *both* scientific *and* social agendas. Contrary to common interpretation, MIT-Masdar is neither a mere episode of technology or practice transfer, nor is MIT in a purely advisory role. MIT's activities furthermore go beyond what is generally called 'technological learning' (Fransman et al. 1984; Kim 1999; Wood & Weigel 2011), a process that to be more unilateral and more specifically targeted at technological (not social institutional) aspects.

For reasons of space constraints, a few selected examples of how MIT facilitated co-productive must suffice. From the beginning, Masdar Institute was conceived and implemented in collaboration with MIT, a partnership funded in full by the Abu Dhabi government through an initial 5-year, $40 million contract (Mills 2009) that was renewed for another 5 years in 2012. As of today, it involved more than 120 faculty, 100 scientists, and 300 graduate students. The partnership is part of a broader current trend towards capacity-building in science, technology, and innovation with the help of foreign 'expert' organizations via complex international collaborations (Pfotenhauer et al. 2016).

Scientifically, first, MIT played a central role in shaping Masdar's original research and education agenda. A group of senior MIT faculty under the leadership of the MIT Technology & Development Program outlined and later concretely defined Masdar's broad research themes and educational programs, the latter modeled closely after MIT's own course offerings in energy and sustainability-related subjects. The process involved specifically those faculty with expertise in energy and sustainability research, gathered in part from the MIT Energy Initiative. While research directions were later adjusted in accordance with the faculty hires undertaken, this facilitation created a temporary, semi-institutionalized set of scientific expertise and a starting point for organizational development.

Second, in terms of recruitment, all initial faculty and administrative positions were advertised and filled under the auspices of MIT, with MIT faculty reaching out to known potential candidates and screening applications for scientific quality and consistency with the research themes. Here, the MIT brand was consciously leveraged as part of the public image of the initiative, granting visibility and scientific credibility to Masdar Institute at an early stage. This gave the fledgling Masdar Institute the desired "international renown" and "caliber of science education" alluded to in the Institute mission (MIST 2011; Masdar Institute 2015).[16] Many Masdar faculty emphasize that the connection to MIT was a key attractor to join the young institution. Similarly, the MIT brand helped to attract high-quality students to Masdar, all of whom receive MIT certificates besides their regular degrees. As a result of aggressive student recruitment, the first round of admissions accepted 89 out of 1200 applicants from around the globe, with an average quantitative GRE score of 765 (comparable to the MIT average).[17] Moreover, several key administrative positions were (temporarily) filled by senior faculty with previous ties to MIT, including the Presidency, the Office of Institute Initiatives, or the Office of Sponsored Programs. Local policy-makers and company leads emphasize how the MIT brand instilled considerable trust in private sector partners and national agencies that Masdar Institute was indeed a credible collaborator and Masdar City a place to set up research operations. By the time of this writing, Masdar has attracted sponsored research centers from Boeing, Etihad, UOP-Honeywell, Safran, IRENA, and UAE's EAD and DSCE, among others.

Third, new Masdar faculty were offered the opportunity to spend their first year at MIT while their labs and other facilities were being constructed in Masdar, which offered some operational slack for Masdar's ramp-up. While at MIT, new faculty were provided with two-year research grants to be shared with an MIT faculty mentor for a collaborative project, contributing to the launching of collaborative projects that could be continued for another year after returning to Masdar and thus establishing an active and sufficiently deep research agenda already *as part of* the start-up process. Conversely, it incentivized MIT faculty to work on Masdar-related issues by funding collaborative projects. As of last year, these faculty-to-faculty collaborations are being increasingly replaced by group-to-group flagship projects. During their visit, most Masdar faculty furthermore audited select courses at MIT with the explicit goal to transfer them to Abu Dhabi. Again, this strong research connection to MIT co-integrated Masdar's nascent scientific activities into an already existing, fully operational and globally visible institutional context.

Fourth, MIT provided an operational 'buffer' for exploring *what exactly* it entailed to run a graduate-level research university in the UAE, for which there was no precedence. MIT helped build a set of social experiences around recruitment processes for faculty, students, and administrators; around questions of supply chains for lab equipment and supplies (including chemicals) as well as maintenance services, all of which usually involves contractors from outside the country; and around cost differences and delays compared to research universities in established science nations, among other things. What is more, many institutional blueprints – from institutional by-laws and IP rules to departmental structure, from educational curricula to salaries and contract templates – had to be created or adopted to Abu Dhabi's conditions. At a larger scale, Masdar Institute was the first to attempt certain infrastructural challenges in the country, again under the auspices of MIT. This included, for example, the creation of first clean room in the UAE and associated downstream architecture challenges, such as hazardous waste treatment. Moreover, many policy frameworks and regulations necessary for advanced research were not in place by the time of Masdar's opening, as they had simply not been relevant in the country up to that point and were 'discovered' only in the context of the MIT-Masdar collaboration. This included, for example, customs regulations for certain pieces of research equipment and hazardous materials, environmental and occupational health and safety (e.g. toxic materials in clean room), or Institutional Review Board (IRB) procedures.[18] All these examples underscore the interwoven nature of scientific and social factors at Masdar Institute, brought to the daylight and recorded as part of the MIT-Masdar partnership. Again, access to MIT as a repository of experience for mature institutional operation and an external site to conduct research before these issues could be fixed contributed to the mutual stabilization of co-production at Masdar.

Finally, the MIT partnership served as an auxiliary institutional support structure for university-based research in an environment where these support structures did not exist. For example, there is no NSF-type independent research funding body. In this context, the MIT-connection provided a quality control mechanism in an environment where research proposals cannot be evaluated for merit through independent peers and in competition with one another. It also provided an additional, collaborative funding mechanism besides governmental block grants to Masdar or private sector funding. At this point, it is not clear whether an NSF-type funding body will be established in Abu Dhabi – however, the need for one has been articulated by both MIT and Masdar Institute participants.

Conclusion

This chapter analyzed Masdar Institute's key role for Abu Dhabi's dual ambition to simultaneously advance Emirati science and society. Technoscientific and social progress in Abu Dhabi, I have shown, are closely interlinked in a co-productionist fashion, whereby each shapes the other in characteristic ways (Fig. 2). Science at Masdar Institute, I have shown, was conceived to usher in a new economic, environmental and social era for the Emirate based on a set of idiosyncratic challenges. To achieve this transformation, Masdar Institute proposed a *particular* type of science – mission-oriented around advanced energy, focused on local relevance, and subject to a comprehensive corporatist-etatist agenda. Conversely, science at Masdar Institute entailed and performed a certain social vision: one of an knowledge-able, scientifically engaged, and aspiring citizenry that manages to balance Emirati identity against a continued reliance on imported people and institutions. I argued that that the co-productionist dynamics at Masdar point to a different emergent settlement of science-society relationships than those typically found in established science nations. This settlement is more transitory and trade-oriented in nature, rooting knowledge production less on permanent institutional allegiance and more on continuous 'caravans' of knowledge traders passing through the country.

The chapter demonstrated further how the MIT-Masdar helped stabilize the very pathways through which Masdar Institute is to unfold its co-productive impact. Masdar Institute drew upon MIT's scientific expertise, operational resources, global visibility and credibility, and physical location as an external site to circumvent and mitigate many co-productionist challenges described above.[19] What is more, the MIT partnership occasionally served as an experimental site – a co-productionist 'probe,' if you will – for discovering the broader social and institutional conditions for building a science system (both upstream and downstream), and subsequently helping to address them. As a result, the partnership was partially able to stabilize the fragile foundations of knowledge production on both the scientific and the social-organizational side. Working with an external partner gave Masdar Institute the necessary support structure to what otherwise would be a highly unstructured, socially unanchored, and open-ended set of activities.

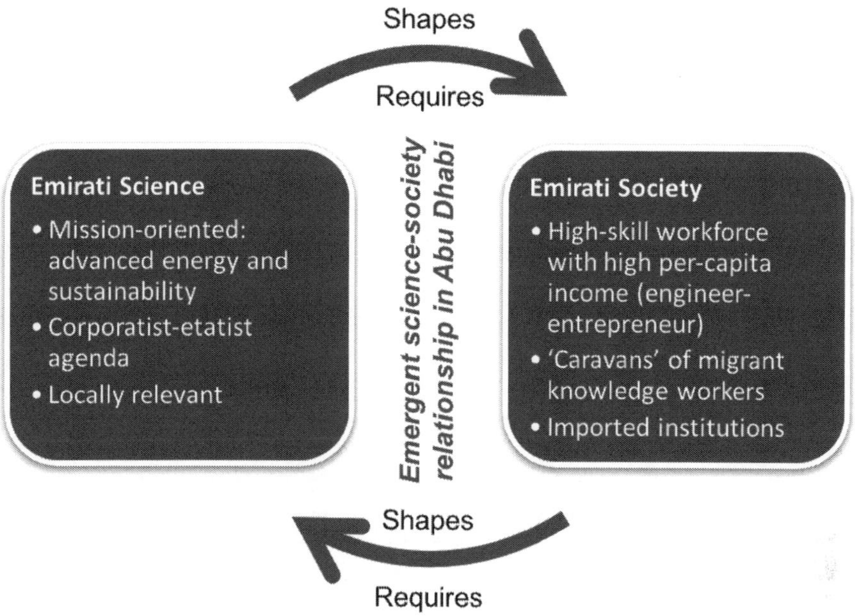

Figure 2: Co-productionist dynamics at Masdar Institute, where both Emirati science and Emirati society are being shaped by their mutual interaction.

The experience at Masdar suggests that international partnerships are a potentially powerful strategy for 'late-developing' countries seeking to rapidly build scientific capacity. We have seen that Abu Dhabi was able to reduce, and at times circumvent, extensive uncertainties and potential development lock-ins by teaming up with an international partner that simultaneously provided scientific, organizational, and operational expertise. Here, the framework of co-production provided a generative lens for understanding the effects of the MIT-Masdar partnership, and offers policy-makers a means to guide and stabilize institution-building processes in fragile contexts.

References

AboulNaga, M.M. & Elsheshtawy, Y.H., 2001. Environmental sustainability assessment of buildings in hot climates: the case of the UAE. *Renewable Energy*, 24(3–4), pp.553–563.

Acemoglu, D. & Robinson, J.A., 2012. *Why nations fail: the origins of power, prosperity, and poverty*, New York: Crown Publishers.

Alderman, L., 2010. Dubai Struggles With Environmental Problems After Growth. *The New York Times*. Available at: http://www.nytimes.com/2010/10/28/business/energy-environment/28dubai.html [Accessed February 7, 2015].

Auty, R.M., 1994. *Patterns of Development: Resources, Policy and Economic Growth*, London ; New York: Hodder Education Publishers.

Beck, U., 1992. *Risk Society: Towards a New Modernity* 1 edition., London; Newbury Park, Calif: SAGE Publications Ltd.

Beck, U. & Grande, E., 2010. Varieties of second modernity: the cosmopolitan turn in social and political theory and research. *The British Journal of Sociology*, 61(3), pp.409–443.

Bound, J., Turner, S. & Walsh, P., 2009. Internationalization of U.S. Doctorate Education. In *Science and Engineering Careers in the United States*. NBER, pp. 59–97.

Braczyk, H.-J., Cooke, P. & Heidenreich, M. eds., 2004. *Regional Innovation Systems: The Role of Governances in a Globalized World* 2nd ed., Routledge.

Bush, V., 1945. *Science, the endless frontier.* 2nd ed., United States Office of Scientific Research and Development.

Butler, J.S. & Gibson, D.V. eds., 2011. *Global Perspectives on Technology Transfer and Commercialization: Building Innovative Ecosystems*, Edward Elgar Pub.

Conceição, P. & Heitor, M., 1999. On the role of the university in the knowledge economy. *Science and Public Policy*, 26(1), pp.37–51.

Corden, W.M. & Neary, J.P., 1982. Booming Sector and De-Industrialisation in a Small Open Economy. *The Economic Journal*, 92(368), pp.825–848.

Croissant, J.L. & Restivo, S., 2001. *Degrees of Compromise: Industrial Interests and Academic Values*, SUNY Press.

Dennis, M.A., 2004. Reconstructing sociotechnical order: Vannear Bush and US science policy. In S. Jasanoff, ed. *States of Knowledge: The co-production of science and social order*. New York: Routledge, pp. 225–253.

Dunning, P.T., 2008. *Crude Democracy: Natural Resource Wealth and Political Regimes*, Cambridge ; New York: Cambridge University Press.

EC, 2012. The European Research Area in the knowledge triangle. Available at: http://ec.europa.eu/research/era/understanding/what/era_in_the_knowledge_triangle_en.htm [Accessed February 7, 2013].

Edquist, C., 1997. *Systems of Innovation: Technologies, Institutions and Organizations*, London: Pinter.

El Din, A.M.S., Aziz, S. & Makkawi, B., 1994. Electricity and water production in the Emirate of Abu Dhabi and its impact on the environment. *Desalination*, 97(1–3), pp.373–388.

Etzkowitz, H., 2002. Incubation Of Incubators: Innovation As A Triple Helix Of University-Industry-Government Networks. *Science and Public Policy*, 29(2), pp.115–128.

Etzkowitz, H., 2008. *The Triple Helix: Industry, University, and Government in Innovation*, New York: Routledge.

Etzkowitz, H. & Leydesdorff, L., 2000. The dynamics of innovation: from National Systems and "Mode 2" to a Triple Helix of university–industry–government relations. *Research Policy*, 29(2), pp.109–123.

Ezrahi, Y., 1990. *The Descent of Icarus: Science and the Transformation of Contemporary Democracy*, Cambridge, Mass: Harvard University Press.

Fichte, I.H., 1971a. *Deducirter Plan einer zu Berlin zu errichtenden, höheren Lehranstalt*, de Gruyter, Berlin.

Foucault, M., 1979. *Discipline & Punish: The Birth of the Prison* 2nd edition., New York: Vintage Books.

Fransman, M., King, K. & Bell, M. eds., 1984. "Learning" and the accumulation of industrial technological capacity in developing countries. In *Technological capability in the Third World.* Macmillan, pp. 187–209.

Freeman, R.B., 2005. *Does Globalization of the Scientific/Engineering Workforce Threaten U.S. Economic Leadership?*, National Bureau of Economic Research. Available at: http://www.nber.org/papers/w11457 [Accessed February 23, 2015].

Freeman, R.B., 2014. *Immigration, International Collaboration, and Innovation: Science and Technology Policy in the Global Economy*, National Bureau of Economic Research. Available at: http://www.nber.org/papers/w20521 [Accessed February 23, 2015].

Galison, P. & Hevly, B., 1992. *Big Science: The growth of large-scale research*, Stanford University Press.

Geiger, R., 1993. *Research and relevant knowldedge: American research univerisities since World War II*, Oxford University Press.

Geiger, R., 1986. *To advance knowledge: The growth of the American research univerisity 1900-1940*, Oxford University Press.

Gelb, A., 1988. *Oil Windfalls: Blessing or Curse?*, New York: World Bank/Oxford University Press.

Gibbons, M. et al., 1994. *The new production of knowledge: The dynamics of science and research in contemporary societies*, Sage Publications.

Giddens, A., 1991. *The Consequences of Modernity* 1 edition., Stanford, Calif.: Stanford University Press.

Government of Abu Dhabi, 2008. *The Abu Dhabi Economic Vision 2030*, Available at: https://www.ecouncil.ae/PublicationsEn/economic-vision-2030-full-versionEn.pdf [Accessed January 31, 2015].

Hausmann, R. & Hidalgo, C.A., 2011. The network structure of economic output. *Journal of Economic Growth*, 16(4), pp.309–342.

Hidalgo, C.A. et al., 2007. The Product Space Conditions the Development of Nations. *Science*, 317(5837), pp.482–487.

Hopwood, D., 2010. Abu Dhabi's Masdar plan takes shape. *Renewable Energy Focus*, 11(1), pp.18–23.

Humboldt, W. von, 2002. *Über die innere und äußere Organisation der höheren wissenschaftlichen Anstalten in Berlin, in: Schriften zur Politik und zum Bildungswesen*, Wissenschaftliche Buchgesellschaft, Darmstadt.

Hunt, J. & Gauthier-Loiselle, M., 2008. *How Much Does Immigration Boost Innovation?*, National Bureau of Economic Research. Available at: http://www.nber.org/papers/w14312 [Accessed February 23, 2015].

Jasanoff, S., 2004a. Ordering Knowledge, ordering society. In S. Jasanoff, ed. *States of Knowledge: The co-production of science and social order*. New York: Routledge, pp. 13–45.

Jasanoff, S. ed., 2004b. *States of Knowledge: The Co-production of Science and the Social Order* 1 New., Routledge.

Jasanoff, S. & Kim, S.-H., 2009. Containing the Atom: Sociotechnical Imaginaries and Nuclear Power in the United States and South Korea. *Minerva*, 47(2), pp.119–146.

Jasanoff, S. & Kim, S.-H. eds., forthcoming. *Dreamscapes of Modernity: Sociotechnical Imaginaries and the Fabrication of Power*, Under review with University of Chicago Press.

Jensen, N. & Wantchekon, L., 2004. Resource Wealth and Political Regimes in Africa. *Comparative Political Studies*, 37(7), pp.816–841.

Jewett, P.A., 2014. *Science, Democracy, and the American University: From the Civil War to the Cold War* 1 edition., Cambridge: Cambridge University Press.

Karl, T.L., 1997. *The Paradox of Plenty: Oil Booms and Petro-States*, Berkeley: University of California Press.

Kim, L., 1999. Building Technological Capability for Industrialization: Analytical Frameworks and Korea's Experience. *Industrial and Corporate Change*, 8(1), pp.111–36.

Krige, J., 2008. *American Hegemony and the Postwar Reconstruction of Science in Europe*, Cambridge, Mass.: The MIT Press.

Landau, A. & Rosenberg, N., 1986. *The Positive Sum Strategy: Harnessing Technology for Economic Growth*, The National Academy Press.

Latour, B., 1993. *We Have Never Been Modern*, Cambridge, Mass: Harvard University Press.
Lundvall, B.-Å., 1992. *National Innovation Systems: Towards a Theory of Innovation and Interactive Learning*, London: Pinter.
Mansfield, E., 1991. Academic research and industrial innovation. *Research Policy*, 20(1), pp.1–12.
Mansfield, E. & Lee, J.-Y., 1996. The modern university: contributor to industrial innovation and recipient of industrial R&D support. *Research Policy*, 25(7), pp.1047–1058.
Masdar Corporate, 2014a. Our Reseach Arm. Available at: http://www.masdar.ae [Accessed January 31, 2015].
Masdar Corporate, 2014b. Who we are. Available at: http://www.masdar.ae [Accessed January 31, 2015].
Masdar Institute, 2015. MIT Partnership. Available at: http://www.masdar.ac.ae/about-us/mit-partnership [Accessed March 9, 2015].
Mercier-Laurent, E., 2011. *Innovation Ecosystems* 1st ed., Wiley-ISTE.
Mills, A., 2009. A New Model for American Colleges Abroad: Quiet Partner. *Chronicle of Higher Education*, 55(29).
MIST, 2011. *Learning to change the world: The Masdar Institute of Science and Technology*, Available at: http://www.masdar.ac.ae/data/p7tp9xg71490.pdf.
Morrison, K.M., 2006. Natural resources, aid, and democratization: A best-case scenario. *Public Choice*, 131(3-4), pp.365–386.
Mowery, D.C. & Sampat, B.N., 2006. Universities in National Innovation Systems. In J. Fagerberg, D. C. Mowery, & R. R. Nelson, eds. *The Oxford Handbook of Innovation*. Oxford, England: Oxford University Press, pp. 1–27. Available at: http://www.oxfordhandbooks.com/oso/public/content/oho_business/9780199286805/oxfordhb-9780199286805-chapter-1.html [Accessed August 24, 2012].
Nader, S., 2009. Paths to a low-carbon economy—The Masdar example. *Energy Procedia*, 1(1), pp.3951–3958.
Neary, J.P. & Wijnbergen, S. van eds., 1986. *Natural Resources and the Macroeconomy* 1st edition., Cambridge, Mass: The MIT Press.
Nelson, R. ed., 1993. *National Innovation Systems. A Comparative Analysis*, New York/Oxford: Oxford University Press.
Nowotny, H., Scott, P. & Gibbons, M.T., 2001. The Co-Evolution of Society and Science. In *Re-Thinking Science: Knowledge and the Public in an Age of Uncertainty*. Cambridge: Polity, pp. 30–49.
Pfotenhauer, S. et al., 2016. Architecting complex international science, technology, and innovation partnerships (CISTIPs): A study of four global MIT collaborations. *Technological Forecasting and Social Change*, 104(3), pp.38–56.
Pfotenhauer, S., 2015. Expertise, legitimacy, and identity: Innovation "best-practices" and the "MIT model" in three start-up universities. In Conference Paper: Can Innovators Be Made? A Dialogue on the Past, Present, and Future of Innovation Expertise. Smithsonian/Virginia Tech.
Pfotenhauer, S. & Jasanoff, S., 2015. Panacea or diagnosis? Imaginaries of innovation and the "MIT model" in three political cultures. *forthcoming*.
Polanyi, M., 1962. The Republic of Science. *Minerva*, 1, pp.54–73.
Ponzini, D., 2011. Large scale development projects and star architecture in the absence of democratic politics: The case of Abu Dhabi, UAE. *Cities*, 28(3), pp.251–259.
Reiche, D., 2010. Renewable Energy Policies in the Gulf countries: A case study of the carbon-neutral "Masdar City" in Abu Dhabi. *Energy Policy*, 38(1), pp.378–382.
Romer, P., 1994. The Origins of Endogenous Growth. *Journal of Economic Perspectives*, 8(1), pp.3–22.

Ross, M.L., 2013. *The Oil Curse: How Petroleum Wealth Shapes the Development of Nations* Reprint edition., Princeton, N.J.: Princeton University Press.
Rudy, W., 1984. *The Universities of Europe, 1100-1914: A History*, Rutherford N.J. : London ; Cranbury, N.J: Fairleigh Dickinson University Press.
Rüegg, W. & Ridder-Symoens, H. de eds., 2003. *A History of the University in Europe: Volumes 1 (Universities in the Middle Ages) & 2 (Universities in Early Modern Europe)*, Cambridge: Cambridge University Press.
Salem, O., 2013. New-look UAE cabinet sworn in by Sheikh Khalifa. *The National.* Available at: http://www.thenational.ae/news/uae-news/new-look-uae-cabinet-sworn-in-by-sheikh-khalifa [Accessed February 2, 2015].
SCAD, 2014. *Statistical Yearbook of Abu Dhabi 2014*, Statistics Centre Abu Dhabi (SCAD). Available at: http://www.scad.ae/SCADDocuments/SYB%202014%20-%20V1%20-%20En.pdf [Accessed November 11, 2014].
Schumpeter, J.A., 1934. *The Theory of Economic Development*, Harvard University Press.
Scott, J.C., 1999. *Seeing Like a State: How Certain Schemes to Improve the Human Condition Have Failed* New edition., Yale University Press.
Solow, R.M., 1956. A contribution to the theory of economic growth. *Quarterly Journal of Economics*, 70(1), pp.65–94.
Stokes, D.E., 1997. *Pasteurs Quadrant: Basic Science and Technological Innovation*, Washington, D.C: Brookings Inst Pr.
Technopolis, 2012. The Evolution of the Knowledge Triangle. *The Technopolitan*, 9. Available at: http://www.technopolis-group.com/resources/downloads/Technopolitan9.pdf [Accessed February 7, 2013].
The World Bank, 2014. *CO2 emissions (metric tons per capita)*, New York, NY: The World Bank. Available at: http://data.worldbank.org/indicator/EN.ATM.CO2E.PC?order=wbapi_data_value_2010%20wbapi_data_value%20wbapi_data_value-first&sort=asc [Accessed February 7, 2015].
Thorp, H.H. & Goldstein, B., 2010. *Engines of Innovation: The Entrepreneurial University in the Twenty-First Century*, Univ of North Carolina Press.
Trow, M., 2002. Reflections on the Transition from Elite to Mass to Universal Access: Forms and Phases of Higher Education in Modern Societies since World WWII. In *International Handbook of Higher Education*. Springer, Dordrecht.
Van de Graaf, T., 2012. How IRENA is reshaping the global energy architecture. *European Energy Review*. Available at: http://www.europeanenergyreview.eu/site/pagina.php?id_mailing=262&toegang=36660e59856b4de58a219bcf4e27eba3&id=3615 [Accessed June 17, 2012].
Varge, A., 1998. *University Research and Regional Innovation*, Boston: Kluwer Academic Publishing.
Weck, O.L. de, Roos, D. & Magee, C.L., 2011. *Engineering Systems: Meeting Human Needs in a Complex Technological World*, The MIT Press.
Winner, L., 1980. Do Artifacts Have Politics? *Daedalus*, 109(1), pp.121–136.
WNA, 2014. *Nuclear Power in the United Arab Emirates*, World Nuclear Association. Available at: http://www.world-nuclear.org/info/Country-Profiles/Countries-T-Z/United-Arab-Emirates/ [Accessed February 7, 2015].
Wood, D. & Weigel, A., 2011. Technological Learning through International Collaboration: Lessons from the Field. In *International Astronautical Congress*. Cape Town, South Africa.
Youtie, J. & Shapira, P., 2008. Building an innovation hub: A case study of the transformation of university roles in regional technological and economic development. *Research Policy*, 37(8), pp.1188–1204.

Notes

1 This work was supported by the NSF Science of Science and Innovation Policy (SciSIP) Program under Collaborative Grant "Technology, Collaboration, Learning: Modeling Complex International Innovation Partnerships" (#1262263), the MIT Leading Technology Policy Initiative, and the MIT-Portugal Program. I would like to thank Fred Moavenzadeh, Sheila Jasanoff, Dava Newman, Dan Roos, Danielle Wood, Mac Hird, the members of the MIT International Advisory Committee, and all interview participants at MIT and in Abu Dhabi for their support and generous sharing of thoughts and information.

2 The figure of co-production shares some common theoretical ground with the notion of "co-evolution society and science" used by Nowotny, Scott, and Gibbons (2001) and Latour's credo of inseparability of nature and culture, the conceptual divorce of which he sees as the foundational myth and futile ambition of modernity (Latour 1993). It is also resonates with the literature on sociotechnical systems (Weck et al. 2011), studies of diverse institutional landscapes of science and systems of innovation (Lundvall 1992; Nelson 1993; Edquist 1997; Braczyk et al. 2004; Varge 1998), as well as research that connects economic growth to social and political institutions (Acemoglu & Robinson 2012).

3 Some of these development challenges reflect what Beck and Grande (2010) call 'compressed stages of modernization.

4 Somewhat surprisingly, they have been scarcely explored from a co-productionist scholarship.

5 While much of this USD 26B urban development project remains to be completed and inhabited (and the original completion date of 2016 will likely be missed), several innovative projects lead the way for this test-bed mission. For example, Masdar City's power infrastructure features a range of renewable energy sources, including a photovoltaic plant, a concentrating solar thermal plant, and a waste-to-energy plant. Another example, an electrically powered personal rapid transport system connects regional transportation hubs outside the city to urban infrastructure and, when fully developed, is foreseen to replace traditional modes of mobility entirely (Nader 2009). What is more, the city has been declared a special economic zone (a policy strategy known from China and elsewhere), rendering resident firms effectively tax-free and eligible for preferential treatment for government services.

6 Originally, Masdar Institute started out with four research thrusts – Sustainable Energy Production and Storage; Efficient Energy Delivery and Use; Water, Environment, and Climate; and Integrated Sustainable Development. The re-grouping was a response to the dual realization that, first, certain sustainability topics cannot be addressed from too narrow a disciplinary basis and, second, sometimes require more fundamental research, particularly into material science and IT.

7 This applies to both the number of Emirati's within the total workforce and average education levels. Of 318,000 Emiratis currently living in Abu Dhabi, about 10.5% hold a university degree (combining both graduate and undergraduate), as opposed to an OECD average of 30%. Emiratis currently make up 15.6% of the local population with university degrees and 10.0% of those with graduate degrees.

8 Educational opportunities for females have also become a trademark of the UAE among its GCC peers, owed largely to the legacy of the late leader Sheikh Zayed, who is still held in high esteem: "Nothing could delight me more than to see a woman taking up her distinctive position in society ... Nothing should hinder her progress ... Like men, women deserve the right to occupy high positions according to their capabilities and qualifications" (MIST 2011).

9 This speculation parallels well-known resource curse arguments around democratic deficits, rent-seeking tendencies, and other governance challenges (Jensen & Wantchekon 2004; Morrison 2006; Dunning 2008; Ross 2013)

10 Currently, 66 countries are represented in the Masdar student body, and 30 in the body of faculty.

11 The list of Abu Dhabi's local universities increasingly reads like an international who's who, including NYU, INSEAD, Sorbonne, and negotiations with Vanderbilt. The Emirate is currently building its own Guggenheim Museum. Moreover, much of the UAE's cityscapes are dominated landmark buildings by star architects (Ponzini 2011). Masdar Institute itself was designed by the London start-architecture

firm Foster & Partners. Abu Dhabi has also attracted the global headquarters of relatively new International Renewable Energy Agency IRENA to Masdar. This has prompted some commentators to point out Abu Dhabi's aggressive "checkbook diplomacy" (Van de Graaf 2012). Similar acts of "checkbook diplomacy" are occurring in other GCC nations, including the world trade rounds and the 2022 Soccer World Cup in Qatar.

12 Other prominent links include the chairman of the board of Mubadala, Masdar's mother company, Sheikh Al Nahyan, Crown Prince of Abu Dhabi and chairman of the Executive Council, the highest organ of government, which ensures that Masdar's concerns are heard at the highest levels. Moreover, Masdar Institute's achievements are regularly interpreted vis-à-vis royal mandates. The educational mission, for example, is almost verbatim quoting the late Sheikh Zayed: "The real asset of any advanced nation is the people, especially the educated ones, and the prosperity and success of the people are measured by the standard of their education" has become somewhat of an official guiding theme of Masdar Institute (MIST 2011).

13 Not all faculty agree with permanent spotlight. One faculty notes that: "I understand Masdar is so new, and [..] we're doing groundbreaking research, if you want to call it like that, [but one] thing that I absolutely hate [..] is that they aggrandize everything. [..] They blow up things out of proportion, where I feel like the role of a scientist becomes less, and the role of a PR person becomes more. [..] For example, you might have run one experiment [..], or maybe have done something unique, okay? But not unique compared to the world. I mean, everyone else is doing this in other labs around the world. They will blow it up. [..] Somebody here bought an instrument. It was a very high tech, expensive instrument, and the media were like all over it [..] It's like he won the Nobel Prize. [..] And they will call someone an expert, when they're not an expert. [..] So for people who are experts in this field, it's an insult, because many of us are junior faculty, so we're not really experts in anything yet. You know, we're starting our career."

14 During his visit in 2013, the author had to "compete" for time and attention with a Nobel Prize winner and two ministerial delegations.

15 Demonstrativeness around scientific and national progress is, of course, not new. STS research has shown many examples where scientific endeavors have come to embody and showcase the functioning of political systems, including representative democracy (Ezrahi 1990); where they have been the political vehicle for, and played out according to, prevailing sociotechnical visions (Jasanoff & Kim 2009; Jasanoff & Kim forthcoming); or where they have called for, or at least been highly compatible with, certain forms of social organization (Winner 1980).

16 One senior leader at Masdar observes: "people recognize that this is an elite university, intellectually [..]. So that is very important for us. You know, we are in the media."

17 Aided, of course, by extremely generous admission packages that include full tuition scholarship, a monthly stipend, free housing, a computer, one return airfare per year to the student's home – and a faculty-to-student ratio of 1:5.

18 Prior to the ramp-up of research at Masdar Institute in the context of the MIT-Masdar partnership, IRB rules in Abu Dhabi had existed only in the form of clinical guidelines, issued by the Health Authority after (a relatively loosely written) federal law, *not* for primarily engineering work with patient data, transportation data, or interview data.

19 Note that I do not wish to suggest that MIT is truly "external" to Masdar in a strict sense. Clearly, this is not the case. What counts as a challenge in Abu Dhabi is shaped by global development, innovation, and sustainability discourses in which MIT plays a central part. Likewise, the envisioned solutions in the form of a research university with associated innovation ecosystem is not independent of MIT, but shaped by both MIT's global reputation and conversations among MIT and Abu Dhabi leadership in the inception of Masdar. Conversely, what is happening at Masdar Institute feeds back into MIT, for example through the sponsoring of research projects at MIT around Abu Dhabi's interests or the re-circulation of education curricula and organizational experiences back to MIT. Research is underway to study these phenomena of circulation.

6

The Research Program at Weill Cornell Qatar: A Unique International Partnership

Khaled Machaca and Javaid Sheikh

Introduction

The term knowledge-based economy is used loosely to describe a broad range of activities that impact economic growth and is built on the concept of innovation-based growth fuelled by national investments in research and development. The OECD (Organisation for Economic Co-operation and Development) defines a knowledge-based economy as "trends in advanced economies towards greater dependence on knowledge, information and high skill levels, and the increasing need for ready access to all of these by the business and public sectors." While Powell and Snellman defines it as: "production and services based on knowledge-intensive activities that contribute to an accelerated pace of technical and scientific advance, as well as rapid obsolescence" (Powell and Snellman 2004). This differentiates knowledge-based economies from more traditional 'resource based'; 'agrarian'; and 'industrial' economies.

At a high level, the rationale for investing in research to help stimulate a knowledge-based economy is simple: investments in research result in the generation of new ideas, solutions, services, skills and intellectual properties that lead to the development of commercial ventures and in turn fuel the economy. Investing in R&D is generally considered to be a key driver of economic growth (OECD 2013), and as one can imagine is a priority for many economies that are currently lopsidedly reliant on natural resources with limited lifetimes.

Of course, for R&D investments to contribute to a knowledge-based economy, there are multiple required building blocks, including but not limited to a robust educational system, comprising K-12 and higher education, geared toward producing well qualified scientists to tackle challenging research questions; well-structured funding mechanisms that direct national resources toward the most promising research; establishment of the physical and logistical infrastructure to support cutting-edge research; a private commercial sector with the knowhow and interest to invest in and develop such technologies and services;

and importantly clear intellectual property, commercial and taxation regulations that would entice investors to support commercialization of the new findings. More generally, research has shown the importance of political and economic institutions that create the right incentives for innovation (Acemoglu and Robinson 2012) Indeed, research on innovation, such as the 'Chain-Linked' model proposed by Stephen Kline argues that new technologies may arise across the continuum of the innovation process from research to the existing body of knowledge or through marketing and design of novel products with each phase continuously feeding on and shaping other phases in the process (Kline 1985, Kline and Rosenberg 1986). Another example would be the 'Circle Model' of innovation outlined by Ralph Gomory, where innovation is achieved by continuously improving on and refining existing technologies that could lead to very significant progress on the long run (Gomory 1989). Nonetheless, although context matters enormously, investments in R&D are a necessary ingredient. The supporting environment that complements innovation typically develops over many decades in countries that have been successful at establishing knowledge-based economies. For young countries in the Arabian Gulf, such as Qatar, with an interest in developing a knowledge-based economy there are significant challenges and opportunities. Qatar has many advantages that support a transition toward a knowledge-based economy, including a visionary committed leadership, significant financial resources, a young population and significant current investments in education and research. However, substantial challenges remain especially given the fast pace at which scientific and technological advancements are occurring globally. The rapid scientific progress and its application favors countries with an established infrastructure and disadvantages those countries that have not yet built it.

Here we discuss a model that Qatar has adopted under the ambitious long-term leadership of the Qatar Foundation, tapping into existing research and education knowhow of leading Universities worldwide to contribute toward the vision of a knowledge-based economy. We will focus on the Biomedical Research Program (BMRP) at Weill Cornell Medical College in Qatar (WCMC-Q), a collaborative effort between the Qatar Foundation and Cornell University that represents a new transnational model that has been put in place to advance education outcomes, as well as research and scientific discovery in Qatar to help contribute to the creation of a knowledge based economy.

An important diversion to discuss before we focus on BMRP program development is the philosophy of national investments in research with the goal of establishing a knowledge-based economy. When trying to promote a knowledge-based economy, some will argue that investments should favor applied research with the goal of showing early successes and a rapid return on investment, thus justifying research investments. Although, this approach has some logic, we would argue that investments in basic research are essential to establish a pipeline of discoveries and knowhow required to fuel a knowledge-based economy on the long term. Although commercialization of such discoveries would be on a longer time frame, investments in basic research can also have an immediate impact on initiating a knowledge-based economy. Such investments create smaller numbers of high technology jobs and training opportunities for locals, and entice the recruitment of foreign talent to create a critical mass of researchers. Furthermore, investments in research encourage the establishment of the service industries that support research efforts such as the supply chain, commercial vendors, specialized contractors, etc. therefore boosting the local economy.

The argument presented above is not meant to dismiss the critical importance of the continuous innovation cycle which shows the interdependence between scientific discoveries and inventions. As discussed by Narayanamurti and colleagues (Narayanamurti, Odumosu, and Vinsel 2013) important discoveries that significantly advance an industry often occur based on previous inventions, (Narayanamurti and Yamamoto 2013). Governmental funding on basic research, however, remains an integral pillar contributing to long-term technology development (2014).

The WCMC-Q Bio-Medical Research Program (BMRP)

In 2001 Cornell University and the Qatar Foundation embarked on a partnership to establish the first medical school in Qatar and the first international branch campus for a US-based medical school. Weill Cornell Medical College in Qatar (WCMC-Q) shares the mission of the main Weill Cornell campus of dedication to excellence in education, patient care and research.

WCMC-Q offers an integrated program based in Doha of pre-medical and medical studies leading to the Cornell University M.D. degree using the same admission, curricular and graduation requirements as those established at the main campus in New York. Clinical teaching and training is conducted in close collaboration with the national health care delivery system in Qatar, Hamad Medical Corporation (HMC).

The WCMC-Q medical school focused on its education mission during its first 8 years of existence since it was founded in 2001. In 2009 the partnership between WCMC-Q and QF was expanded to include the establishment of a biomedical research program, in conjunction with a national drive in Qatar to invest in and develop the country's research infrastructure an effort championed and led by the Qatar Foundation. Concurrently, QF established an entire ecosystem in Qatar to empower the research mission toward a knowledge-based economy. First, multiple leading US Universities under the umbrella of Education City where invited to establish branch campuses in Qatar, creating a unique multi-university model that provides a cutting edge education for local students. The different institutions with established branch campuses in Doha include Texas A&M, Carnegie Mellon, Georgetown, Virginia Commonwealth, and Northwestern.

Second, QF established a national funding agency, the Qatar National Research Fund (QNRF), which has been extremely successful in building peer-reviewed competitive funding mechanisms that are currently sought after by researchers in Qatar and internationally. Finally, QF developed the Qatar Science & Technology Park (QSTP) a technology incubator with a mission of developing new findings and intellectual property from the different research institutions in Qatar into commercially successful entities. This infrastructure is critical to support research efforts and translate them in viable commercial entities especially for a young research ecosystem as in the case in Qatar. The funding agency QNRF invests in the best research in the country based on a robust peer review system that rightfully so would use national resources to build a research portfolio and culture in Qatar. Research funded by QNRF is conducted at different academic and national research centers and the findings, it is hoped, will be managed and commercialized by the technology incubator QSTP.

More recently, starting in 2010, QF developed three national institutes to support the research effort in the country, with a focus on computing, energy and the environment, and biomedical sciences. Finally an Emiri Decree was issued in 2010 by his Highness the father Emir Sheikh Hamad bin Khalifa Al-Thani dedicating 2.8% of the GDP in Qatar to research (Group 2011), one of the highest in the world and the highest in the region. It is within this comprehensive, young, supportive ecosystem that the BMRP program saw its beginnings in 2009.

Mission & Vision

The mission statement of the BMRP program is to:

> *"Establish a center of excellence in basic, translational and clinical research*
> *with a world-class infrastructure, and outstanding scientists*
> *to target the most pressing health needs in Qatar and the region."*

When one is faced with the task of establishing a new research program, it offers a distinctive opportunity. The BMRP program was designed by the Weill Cornell and QF leaderships to focus on the most pressing health needs for Qatar as a nation. It was readily apparent that the most significant health challenge for the country is the metabolic syndrome (diabetes, obesity and the associated cardiovascular disorders) epidemic. With an estimated diabetes prevalence of 16.7% (Bener et al. 2009), and staggering estimates on the prevalence of obesity of 72% of the population being overweight (BMI >25kg/m^2) and 33% being obese (>30kg/m^2) (Bakri and Al-Thani 2013, Rahim et al. 2014), it was clear that this was a healthcare issue that could not be overlooked for Qatar. This trend also applies to other countries in the region. Furthermore the Middle East is expected to have the highest rate of increase in diabetes in the world estimated at 164% from 2000-2030 (Hossain, Kawar, and El Nahas 2007). Neurogenetic disorders were also of interest given the depth of knowledge available at Weill Cornell and their prevalence in Qatar. Therefore, the vision for the BMRP program was to focus, at a high level, on metabolic syndrome and neurogenetic disorders with the ultimate goal of making an impact on these diseases and their management nationally and regionally to impact the efficiency of health care delivery, and usher in the era of precision medicine in Qatar. Precision medicine is a novel concept that postulates devising targeted clinical treatment for patients individually or as small groups based on their molecular phenotype. Currently medical approach to a particular disease use standardized treatment modalities that are approved and well validated. Although these treatments work for the majority of the patients they sometimes are not ideally suited for a subset of patients as they ignore individual variations that may mitigate the effectiveness of the treatment. A good example to illustrate the point is the case of non-small cell lung cancer (NSCLC) where 5% of the patients were found to carry a specific mutation that activates a specific gene (the ALK kinase). This subset of patient did not respond well to the standard treatment, however when treated with ALK inhibitors the success of the treatment improved significantly (Villaruz et al. 2013). This illustrates the potential power of precision medicine approaches to improve the quality of health care delivery for individual patients while reducing the inherent costs.

Approach

Establishing a biomedical research program in a young country such as Qatar presented many infrastructural and logistical challenges that the research team at WCMC-Q was able to overcome only through close coordination with and support from multiple local stakeholders. Most importantly the support from Qatar Foundation has been instrumental.

Tackling complex diseases such as diabetes and obesity is a daunting challenge that will require a global effort. For the BMRP to make a local impact in that arena, we chose to focus our efforts on factors that may differentially affect the local population in Qatar. Particularly genetics/genomics factors that are likely to give us unique insights into these complex diseases given the high rates of consanguineous marriages in Qatar. Our goal is to tackle such diseases from the basic molecular and cellular mechanisms all the way through translational and clinical factors.

A summary of the vision for the BMRP to build research strength that spans the basic, translational, clinical and public health arenas is illustrated in Figure 1. Briefly the goal of WCMC-Q was to establish the physical, administrative and human capacity locally in Qatar to address complex diseases to impact healthcare in the country. The leadership team, in which we are members, viewed the interaction among the different levels of research as continuously feeding back and informing each other to define next steps in the research effort. Finally from the inception of the program, a set of deliverables and timeframes were identified to keep the program focused. These deliverables are outlined in Figure 1 and will be discussed in further detail below. Such an approach is likely to improve our understanding of disease factors specific to the local population, impact the efficiency and adaptability of health care delivery, and usher in the era of personalized medicine in Qatar because it will provide novel molecular information in the context of the diseases of interest for the local and regional populations.

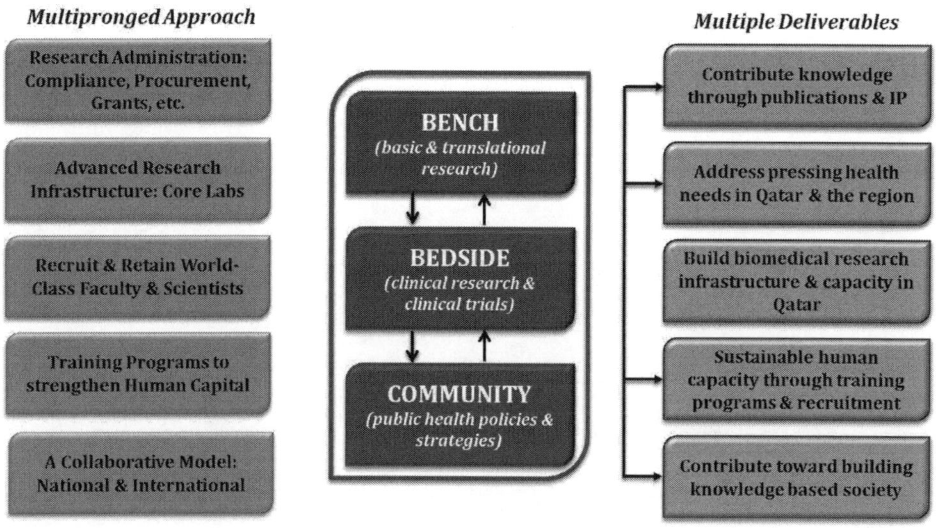

Figure 1: BMRP Vision. Developed by the authors.

Program Structure

As illustrated in Figure 2, the research program at WCMC-Q, which is comprised of faculty, postdocs, and technicians, is structured into three inter-related categories: 1) Research Administration; 2) Centralized Research Core Facilities; and 3) Faculty Labs with broad based expertise focusing on understanding diseases prevalent in the Qatari population.

Research Administration

A. Approach to Program Development

The establishment of the administrative infrastructure to support the BMRP research programs is a fundamental component of the program. Successful program development requires experience from seasoned professionals with a history of sustained accomplishments. The administration's management philosophy is to formulate and clearly communicate the program's overall vision, set achievable goals and objectives, combine individual skills and experience, and provide support to nurture and develop human capital.

Our management structure is built with the aim to provide maximum adaptability to respond to change and emerging opportunities, which is required in any research environment, but is especially important for program development in a rapidly growing country such as Qatar.

WCMC-Q research administration is divided into three departments: business and grants office; office of research compliance; and a laboratory management component.

Our approach to establishing our administrative structure was to design support services that fit local needs. We have developed workflows that comply with institutional and local policies to meet the specific needs and challenges that faced program development. These workflows were aimed to empower local researchers to focus on their science while supporting them with critical functionalities such as procurement and recruitment of research personnel.

Figure 2: BMRP Structure. Developed by the authors.

One example of this approach is the development of 'job families' that fit the local culture and needs to allow us to recruit the best local talent. The largest group of employees in Research is the research specialists, who are recruited locally. The equivalent title for these positions in the USA is 'research technician'. However, in the context of the local culture the title of "technician" is viewed with dissatisfaction by the employees. We therefore created the title of "Research Specialist" to better reflect the broad range of skills, education and experience that we are expecting and were fortunate to have in our local hiring pool. To help alleviate retention problems and provide local hires on the Research Specialist track with a clear career path, we implemented a progression of 5 levels one can achieve in this category of employment based on merit and productivity. This approach has been quite successful at reducing turnover and improving employee morale. So far over 100 Research Specialists have been trained at WCMC-Q with ~70 currently employed in WCMC-Q Laboratories.

B. Business & Grants Office

The business and grants office, handles recruitment and management of research personnel, development of new research outreach programs, administration of existing programs, faculty support with grant proposal, management of contracts and assurance of compliance with sponsor requirements. Staff in the Business and Grants Office support faculty in the financial management and submission of their proposals primarily to the Qatar National Research Fund (QNRF), and procurement of reagents and equipment that are essential for research conduct.

C. Research & Ethical Compliance

In keeping with the general approach of the BMRP, the Office of Research Compliance has created Qatar-based compliance structures; participated in the national dialogue about research compliance; and collaborated with other institutions in building national research compliance capacity.

WCMC-Q has taken a phased approach to research compliance, just as it has with other aspects of research administration. This approach involves leveraging WCMC-NY expertise and experience, and then phasing in local structures as WCMC-Q research expands in terms of volume and complexity. A major achievement using this approach has been the establishment of a local Institutional Review Board (IRB) to review and oversee research involving human volunteers. This local IRB has evolved into a Joint-IRB (JIRB) with our partner institution Hamad Medical Corporation (HMC), which facilitates the review process for joint projects and avoid significant duplicative administrative efforts. The JIRB brings together members with a wide range of expertise and from many different backgrounds. In addition to HMC and WCMC-Q faculty, the IRB membership has included members from Qatar University (QU), the Faculty of Islamic Studies and members of the community. This diversity of membership provides the IRB with a wealth of ethical perspectives about the local research context and helps integrate WCMC-Q with its partner institutions.

IRB members and compliance staff have been active voices in the national dialogue about research compliance. IRB members serve on the Supreme Council of Health's (SCH) Qatar National Research Ethics Committee. Compliance staff and IRB members regularly provide feedback to the SCH regarding draft policies and participate in SCH-organized events.

D. Laboratory Management

Our research space employs the modern open Lab format, which greatly facilitates space and personnel management. Reliable management of this open space involves multiple functionalities, including for example maintenance of common laboratory areas and equipment, upkeep of service contracts on equipment, coordinating with facilities to maintain the day to day operations of the laboratories, and stocking of common supplies. The laboratory management personnel ensure that all these functionalities are operational to facilitate scientific productivity.

Core Laboratories

Core Labs provide centralized functionalities and platforms that are not realistically achievable in individual laboratories. These central core labs provide an essential support infrastructure for the advancement of biomedical research in Qatar. With the establishment of these various services, WCMC-Q has created a modern research support infrastructure, one that empowers existing scientists and that is attractive to potential recruits. Many of the cores described below depend on advanced platforms to provide their services. Conceptually the cores at WCMC-Q, in addition to providing a specific service for faculty, also conduct their own research projects and train staff, thus contributing to the overall mission of the program of building human capital and improving scientific discovery. The specific technologies for the core labs were careful chosen to support and drive the scientific needs of the program. All of the established cores are currently used at capacity and have resulted in significant collaborative projects among our faculty.

A. Genomics Core

The genomics core is a state of the art facility that provides functionalities covering the full range of modern genomics analyses including: PCR, sequencing, SNP mapping, microarray technology for gene expression analysis, library construction, genotyping, and 2nd and 3rd generation sequencing capabilities.

B. Imaging Core

The imaging core provides advanced functionalities for cellular imaging, including a deconvolution system and fully spectral confocal microscopes that will enable cutting-edge cellular imaging approaches, which are becoming central to biomedical research. The imaging core also encompasses a flow-cytometry and cell sorting facility and histology services.

C. Basic Core

The basic core provides basic functionalities that are essential for practically every lab. Those include but are not limited to cell culture facilities, shakers, incubators, gel documentation, standard and high-throughput plate readers, spectrofluorometers, etc.

D. Proteomics Core

The proteomics Core focuses on quantitative proteomic approaches and post-translational modifications of proteins. The Core uses top of the line, high resolution and high mass accuracy mass spectrometry, and offers identification and relative quantification of proteins in complex mixtures. Depending on demand capabilities in the portfolio are set to grow in the direction of large scale analysis of post-translational modification and quantitation strategies based on chemical labelling.

E. Bioinformatics Core

The bioinformatics core supports PIs in analysis of high throughput genomics and proteomics data. In this era of systems biology such a core is essential to analyze and mine high throughput data effectively. In addition the core will cater to additional computational needs to analyze biomedical data for individual projects. The core also assists users to access different supercomputing facility, both in Qatar and at the Cornell main campus. As a sub-division of the Bioinformatics Core, the Virtual Metabolomics Core assists investigators in using metabolomics for their research.

F. Biostatistics Core

The Biostatistics and Biomathematics Research Core provides services and expertise to assist investigators within WCMC-Q and other institutions in Qatar in all fields of biostatistics and epidemiology, including study design, data analysis, and grant proposal preparation. The Core is equipped with three high-computing workstations that run multiple statistical, mathematical, and modelling specialty software. The Core assists investigators in study design for clinical trials, cohort studies, case-control studies, validation studies or any other observational studies. The Core also performs appropriate sample size calculations and power analyses. In addition, the Core provides comprehensive statistical data analysis, access to Core computing facilities, and assistance in output interpretation.

G. Clinical Research Support Core

An important component for success in translational research is the establishment of adaptable and sustainable clinical research programs in collaboration with local hospital systems. For clinical research support to be successful it needs to evolve to meet the needs of the dynamic evolving research environment in Doha that comes with its set of challenges. The Clinical Research Support Core provides multiple services to support PIs in their clinical studies including training for coordinators, a reliable central sample repository, and an electronic clinical data management system that will allow integration of phenotypic data with sample tracking. On the long term the goal is to combine additional 'omics' data into a single searchable relational database that will facilitate meta-analyses. Such a dataset will provide a rich resource that clinical investigators in Qatar can employ to ask specific clinical questions.

Recruitment and Establishment of Faculty Laboratories

Nearly 160 scientists and staff members are currently actively involved in the WCMC-Q research program, including research faculty, postdoctoral fellows and research associates, and research specialists. The growth in personnel directly employed by the BMRP is illustrated in Figure 3. The initiation of the research program at WCMC-Q has led to the involvement of a large number of staff and faculty from other divisions within the institution. The exponential growth in personnel involved in research over the past 5 years attests to the success of the program in meeting one of its primary goals of human capacity building through recruitment and training programs. The personnel growth was coupled to setting the proper infrastructure capable of supporting and sustaining research conducted by the teams recruited to and trained in Qatar. This section describes how in a relatively short time the BMRP has been able to develop a substantial and dynamic research community.

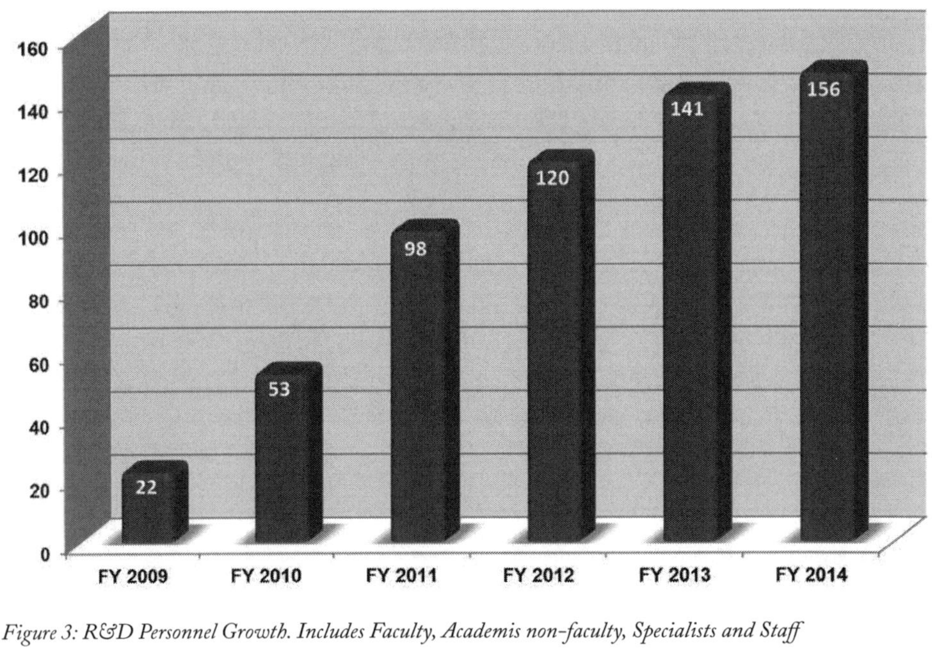

Figure 3: R&D Personnel Growth. Includes Faculty, Academis non-faculty, Specialists and Staff

A. Recruitment of Research Faculty

The backbone of the research program consists of the individual faculty laboratories that are tackling different biological problems to ultimately generate the knowhow that will improve the treatment and prognosis of diseases relevant to the Qatari population. The research administration infrastructure and core Labs support individual faculty efforts to advance their scientific program. Given the vital role of individual research faculty in the success of the overall program, the selection of the best-qualified candidates to lead faculty laboratories has been rigorous. Important considerations that were taken into account for Faculty recruitment, include a technology transfer component to Qatar, a focus on research areas of importance to the BMRP program as discussed above, and a track record of high caliber research productivity and potential. Initial recruitment efforts concentrated on establishing broad scientific expertise in Qatar to address diseases of importance for the Qatari population and the region through personalized medicine approaches. The specific foci were on diabetes, obesity, metabolic syndrome and neurogenetics as the most pressing health need for Qatar. A broad-based recruitment committee with membership from QF, WCMC-Q and WCMC was established and cast a broad net internationally through advertisements and direct communications to attract the best possible faculty to Qatar. This effort has been quite successful as it resulted in the recruitment of both junior and senior faculty from premier institutions around the world such as the Max Planck, Harvard, and Columbia.

B. Recruitment of Research Specialists and Non-Faculty Academic Staff

An essential component of establishing a biomedical research program is the recruitment of qualified scientific talent to the Laboratory. In the absence of a graduate school at WCMC-Q, the laboratories are staffed with research specialists who are recruited and trained locally, and two types of non-faculty academics: Postdoctoral Fellows and Research Associates. Research specialists are discussed in more details in the training and capacity building section. Postdoctoral fellows and research associates are recruited internationally to provide the specialized expertise for individual research Labs. Non-faculty academics are typically recruited on multi-year contracts and many of trainees recruited to WCMC-Q have chosen to stay in Qatar either by moving to other institutions or progressing within WCMC-Q. Just like for faculty academics, recruitment of non-faculty academics was challenging at the initiation stages of the program because non-faculty academics were understandably concerned about their ability to be productive in the young research environment being established at WCMC-Q. However, currently recruitment of non-faculty academics had improved significantly with Labs consistently able to recruit to talent because previous fellows have been quite successful at publishing papers in high impact journals.

Sustainable Human Capacity

Human development is a pillar of the Qatar National Vision 2030 and an integral part of the BMRP. To build sustainable domestic research capacity and develop a true research community, it is necessary to train local young talent and engage those who are interested through outreach programs. As detailed below, over the past years we have created various new programs, under BMRP, which have contributed to the development of a robust and expert research workforce. This training is already contributing to the growth of biomedical research in Qatar. In addition to summer research training programs targeted to our medical and pre-medical student populations we have established and contributed to other training modules in Qatar. Given the importance of this training to the human capacity building component they are briefly discussed below.

A. Qatar Science Leadership Program (QSLP)

The QSLP is a training program managed by the research division at Qatar Foundation. WCMC-Q is one of the institutions taking part in the program. QSLP identifies exceptional young Qatari scientists and supports their further training and education. WCMC-Q has hosted close to 20 QSLP trainees and provided them with training in both research administration and hands on experience in the Labs. QSLP candidates may choose to continue this basic training by enrolling in a PhD program. Currently several QSLP trainees are undertaking their PhD thesis research in WCMC-Q research Labs. Those trainees who chose the Administration track are using skills acquired during their time at WCMC-Q in new positions at various institutions in Qatar.

B. Research Specialists & Clinical Coordinators

An important aspect of instituting a sustainable research infrastructure is the training of the local workforce in cutting-edge scientific technologies and approaches. Toward that goal WCMC-Q has over the past 5 years recruited and trained locally over 100 research specialists and clinical coordinators. These scientists are graduates of local and regional universities, primarily Qatar University. This is an on the job training program that offers candidates a position in a research Labs at WCMC-Q coupled to training in specialized techniques that matches the Lab's expertise. In addition to training lab staff, it became immediately clear that employee satisfaction and retention are critical factors to consider for these positions. This effort is supported by the human resources division, which hosts regular workshops aimed at engaging employees and helping them refine their performance and further define their careers opportunities. Those include topics such as: interviewing skills, e-learning, inspiring leadership, performance management, cultural diversity, workplace behavior and diversity. WCMC-Q is also increasing the number of programs offered specifically to research division employees to encourage team-building and collaborations in a rapidly expanding workplace.

C. WCMC-Q Training Program for Qatari Nationals

BMRP instituted an internal training program for Qatari Nationals in an effort to increase the number of nationals engaged in biomedical research and enhance the representation of Qatari employees at WCMC-Q. This program is a comprehensive training program that exposes students to both research administration and basic or clinical research based on the trainee's interest. The training is 6 months long and includes a stipend and a completion certificate for each trainee. During that time the trainees are employed by WCMC-Q on a full-time basis. Candidates who hold full-time employment before joining the program have been able to negotiate a leave of absence from their employer to engage in this training program while maintaining their employment. This highlights the flexibility and contribution of employers in Qatar to support the training of their staff and investment in continuing education while on the job. The national training program has proven to be exceptionally successful and has attracted the full spectrum of students from fresh graduates to physicians who have completed their fellowship training. One of the program alumni has chosen to join the medical program at WCMC-Q. Importantly the program is already contributing toward increasing the numbers of nationals engaged in biomedical research both administratively and at the bench, and also increasing the population of Qatari nationals employed at WCMC-Q. In that context currently 27% of the medical and premedical student body at WCMC-Q is represented by Qatari Nationals. Furthermore, Qatari students represent the largest single nationality in the last admitted medical student class. The authors are currently studying the potential and feasibility of research training programs targeted at both high school and college Qatari students, with the hope of developing broad spectrum programs covering high school through early-career professionals--all aimed at helping young adults best refine their career aspirations and targeting opportunities for their career development inside Qatar.

Research Outcomes

To monitor the productivity and success of the BMRP, we have adopted a set of key performance indicators. Some are typical for any research program, such as publications/patents and extramural funding, whereas others such as human capacity development and local collaborations are more specifically targeted toward the BMRP mission in Qatar.

A. Publications and Patents

The caliber of scientific peer reviewed publications represents the gold standard by which the success of a research program is judged. New knowledge is contributed in the form of scientific publications and intellectual property. Since BMRP inception, WCMC-Q faculty have published over 330 peer reviewed papers, some in the most prominent Journals in the biomedical sciences. This reflects the high quality of the research and without question "puts Qatar on the map" in the biomedical arena. In addition WCMC-Q faculty filed 3 patents, one in collaboration with Qatar University, highlighting the start of our intellectual property portfolio.

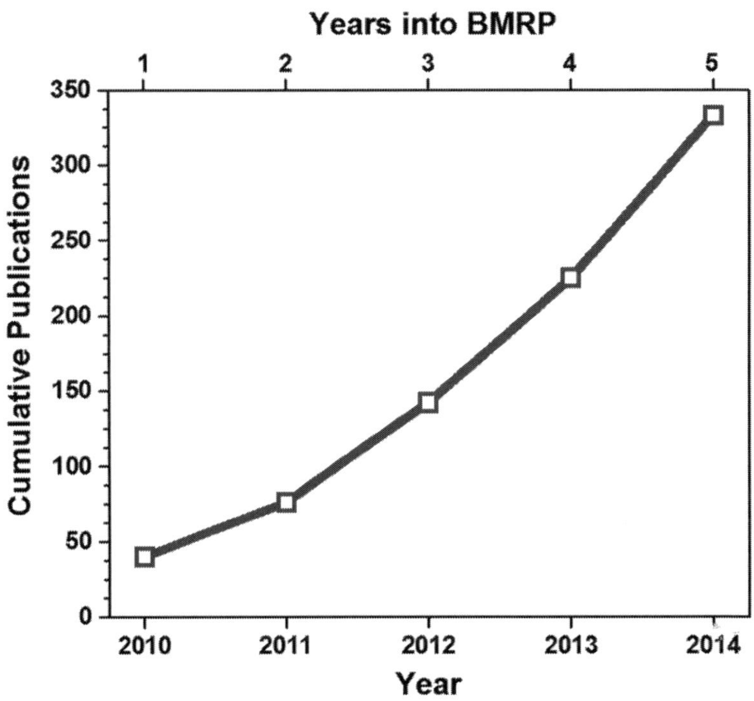

Figure 4: BMRP Publications.

B. Extramural Funding

Extramural funding through competitive peer-reviewed grants as offered by Qatar National Research Fund (QNRF), the National funding agency in Qatar, is the engine that drives creative research. WCMC-Q faculty received $90.2 million in cumulative competitive funding from QNRF since 2009.

C. Collaborations with Local Stakeholders

Our philosophy for BMRP development is centered around a collaborative approach with local stakeholders engaged in biomedical research, including the hospitals in Qatar. Some examples of these collaborations include: Grants teaching and training collaborations with HMC and Qatar University; Close collaborative effort with the Supreme Council of Health to establish the compliance infrastructure in the country; Outreach and collaborative research projects with the Qatar Diabetes Association, to name a few.

D. Human Capacity

As noted above, in just 5 years we grew from a handful of staff and faculty involved in research to a current team of 37 research active faculty, 44 non-faculty academics, 66 research specialists, and 21 administrative staff including those involved in ethical compliance. This growth in human capacity provides the critical mass of scientists and support staff to maintain a sustainable research program.

Conferences and Workshops

Access to a vibrant research community through conferences represents an excellent venue to support capacity building in Qatar and encourage collaborations. WCMC-Q hosted several such events. *(1)* WCMC-Q Research Retreats - hosted annually for the past three years (FY11, 12 and 13) to highlight the achievements and progress made by our scientists and students. *(2)* Departmental Seminars - more than 160 since 2009. *(3)* Five International Conferences: The DALM (Drugs Affecting Lipid Metabolism) a major international conference first established in 1960 made its debut in Doha in 2011 with a focus on Metabolic Syndrome, with more than 400 scientists attending. Additional conferences include the "Neurogenetics Workshop" (April 2011), "Psychiatry Think Tank" (April 2012), "Spina Bifida" (Sept 2013) and "Neurosciences" (planned for March 2014). *(4)* Eleven additional local conferences/workshops were organized from 2010 to date, focusing on diverse areas, including, among others, stem cells, cardiovascular, neurogenetics, and informed consent.

Conclusion

Over the past 5 years the BMRP has established the technical, human and logistical infrastructure to operate and grow a productive biomedical research program in Qatar. This infrastructure represents a resource for the biomedical community in Qatar and the region, and promises significant advancement in the context of the vision toward a knowledge-based economy through partnerships among local stakeholders.

The Future

Given the robust infrastructure put in place the future of the BMRP is promising. In the next phase we will focus our efforts along two primary lines. First, the development of a graduate school in the biomedical sciences in collaboration with local stakeholders, including Hamad bin Khalifa University (HBKU) and the Qatar Biomedical Research Institute. Graduate training is a critical aspect of a successful academic research program, which is currently not a part of BMRP. It infuses the program with a vibrant, intellectually eager young workforce. It also keeps faculty engaged in training and graduate teaching. In Qatar there is a pressing need for graduate degree offerings in the biomedical sciences. This is because the only university in Qatar currently offering graduate degrees in biomedical sciences is Qatar University (QU), and the demand for graduate studies vastly exceeds QU's current program capacity. Furthermore, the need for diversity in program offerings enhances the demand for additional graduate programs in the biomedical sciences in Qatar. WCMC-Q collaborates with QU in terms of hosting graduate students and joint faculty appointments. These partnerships are important as they encourage scientific collaboration among local stakeholders. A degree granting graduate program would offer students the ability to earn a terminal degree in the biomedical sciences from an Ivy League institution in Qatar. This is important to involve Qatari students, given the cultural pressures many of them are under to stay close to family during their studies. Furthermore the Qatar National Research Strategy (QNRS) aims at dramatically increasing the critical mass of scientists in Qatar to support the National vision of establishing research center of excellence. Hence, a local graduate program to provide talent and support the expected exponential growth in the number of scientists in Qatar becomes a timely and essential strategic move. Furthermore, a high level of Qatari students' enrollment in these graduate programs will be essential to achieve. This will require a consolidated broad spectrum approach to inform Qatari college students about career options in Qatar for MS and PhD biomedical research graduates. Furthermore, research training and research internship programs for college students become critical to provide with early exposure to research as a potential career option. Based on our experience with our medical program in Qatar it would also be advisable to extend these outreach programs to high school students to get engaged into the research endeavor early on. The national funding agency, QNRF, has already recognized these challenges and has instituted funding mechanisms for both high school and college students to get engaged in research. Furthermore, QNRF has established a funding mechanism for graduate studies as well. As such the stage is already set nationally to build on these funding mechanisms toward establishing successful graduate program in Qatar.

Finally, an important goal that we are already making significant progress on as discussed above is increasing the engagement of Qatari students in biomedical research in general and as scientists and administrators at WCMC-Q specifically. It is important to consider the long term sustainability of the ongoing research programs in the context of benefiting Qatar and showing a significant return on investment nationally. Qatar's vision is to establish a knowledge-based economy where national investments in research will produce a pipeline that will lead to a self-sustainable commercialization effort. Establishment of young companies will in turn close the cycle and result in new jobs and financial returns for the country. For this model to be maintainable in Qatar, it will be important to establish the local knowhow and leadership that would oversee and drive this cycle. Hence investment in the training and promotion of young Qatari talent must be at the forefront of research efforts. The Nationals training program discussed above is already contributing in a small fashion toward providing the infrastructure to achieve this goal.

Key Success Factors

What are the ingredients required for a young research program to establish itself in the Gulf and successfully produce scholarship and training? Our experience developing the BMRP in Qatar provides some insights outlined below in bullet point format:

- Strong support from the national leadership is critical.
- Organizational structure and administration must be adapted to local needs and culture.
- Broad-based collaborations among local stakeholders are essential.
- Flexibility and adaptability are key ingredients to success. Neither WCMC-Q nor the Qatar Foundation could have predicted the hurdles that faced the development of a biomedical research program in Doha as it had never been attempted before. The flexibility to tackle logistical problems is critical to support scientific progress, which is the most fundamental deliverable.
- Recruitment of outstanding faculty to Qatar is achievable when the proper infrastructure is in place.
- There exists a substantial pool of local young scientists looking for real opportunities to grow professionally in an intellectually stimulating environment.
- Establishing and supporting cutting edge platforms, (Core Labs), greatly enhances the research environment and collective scientific capabilities.
- Training and outreach are vital to generating sustainable human capital and enhancing the transfer of technology and knowhow to Qatar.

References

2014. RESTORING THE FOUNDATION: The Vital Role of Research in Preserving the American Dream. Cambridge, Massachusetts.

Acemoglu, Daron, and James A. Robinson. 2012. *Why nations fail : the origins of power, prosperity and poverty.* 1st ed. New York: Crown Publishers.

Bakri, H., and A. Al-Thani. 2013. Qatar STEPwise Report 2012: chronic disease risk factor surveillance. Supreme Council of Health.

Bener, A., M. Zirie, I. M. Janahi, A. O. Al Hamaq, M. Musallam, and N. J. Wareham. 2009. "Prevalence of diagnosed and undiagnosed diabetes mellitus and its risk factors in a population-based study of Qatar." *Diabetes Res Clin Pract.* 84:99-106.

Gomory, R. E. 1989. "From the 'ladder of science' to the product development cycle. Scout for technology, design for manufacture, shorten the time to market." *Harv Bus Rev* 67 (6):99-105.

Group, Oxford Business. 2011. *The Report, Qatar 2011.*

Hossain, P., B. Kawar, and M. El Nahas. 2007. "Obesity and diabetes in the developing world--a growing challenge." *N Engl J Med* 356 (3):213-5. doi: 10.1056/NEJMp068177.

Kline, S. 1985. "Innovation is not a linear process." *Research Management* 28 (4):36-45.

Kline, S. J., and N. Rosenberg. 1986. "An overview of innovation." In *The positive sum strategy: harnessing technology for economic growth*, edited by Ralph Landau and Nathan Rosenberg, 275-305.

Narayanamurti, V., T. Odumosu, and L. Vinsel. 2013. "RIP: The Basic/Applied Research Dichotomy." *Issues in Science and Technology* 29 (2):31-36.

Narayanamurti, V., and K. Yamamoto. 2013. ARISE II: Unleashing America's Research & Innovation Enterprise

OECD. 2013. "Raising the Returns to Innovation: Structural Policies for a Knowledge-based Economy. In *OECD Economics Department Policy Notes.*

Powell, W. W., and K. Snellman. 2004. "The knowledge economy." *Annual Review of Sociology* 30:199-220. doi: DOI 10.1146/annurev.soc.29.010202.100037.

Rahim, H. F., A. Sibai, Y. Khader, N. Hwalla, I. Fadhil, H. Alsiyabi, A. Mataria, S. Mendis, A. H. Mokdad, and A. Husseini. 2014. "Non-communicable diseases in the Arab world." *Lancet* 383 (9914):356-67. doi: 10.1016/S0140-6736(13)62383-1.

Villaruz, L. C., T. F. Burns, V. S. Ramfidis, and M. A. Socinski. 2013. "Personalizing therapy in advanced non-small cell lung cancer." *Semin Respir Crit Care Med* 34 (6):822-36. doi: 10.1055/s-0033-1358552.

7

The Royal College of Surgeons in Ireland Bahrain: Changing the Medical Landscape in the Middle East Through International Collaboration and Knowledge Transfer

David Whitford, Cameron Mirza and Annamarie Lawrence

Introduction

The changes within the higher education landscape of the countries of the Arabian Gulf in the past thirty years are striking. With many of the Gulf Cooperation Council (GCC) countries stating their desire to develop a knowledge-based economy and move away from a reliance on dwindling energy resources, there has been a significant focus on the role education can play in aiding this development, especially in the areas of science, technology and innovation. Of the various waves of educational reforms in the Gulf, internationalization has been one of the most significant (Knight 2013, Donn and AlManthri 2010, Wildavsky 2010) One particular aspect of internationalization is the establishment of international branch campuses (Lane 2011). Western higher education institutions have responded to the invitation and incentives of the GCC governments to set up campuses as key knowledge centers within GCC countries. The incentives for the creation of these knowledge centers have been complemented by policies and strategies aimed at developing local work force capability with the desire to minimize reliance on expatriate labor. This strategy has seen the emergence of 'education hubs' within some of the GCC countries, namely Qatar and the United Arab Emirates (UAE), as a collection of branch campuses of prestigious international universities to compete for government support and student enrolments. Education City in Qatar has attracted five prestigious American universities including Weill Cornell Medical College. Dubai International Academic City in the UAE is home to 27 different institutions from 10 countries. Limited by lower levels of wealth than some of its neighbors, the Kingdom of Bahrain has taken a slightly different approach to attracting international branch campuses by focusing their education reforms on building local capacity as a priority over importing expertise (Higher Education Council 2014a).

As we discuss below, Bahrain has retained a higher percentage of the nation's jobs for the national population and has encouraged high levels of enrolment of country nationals in higher education (Donn and AlManthri 2010, Karolak 2012). However, Bahrain is lagging behind some of its neighbors in developing capacity in science, technology and innovation (STI) (Higher Education Council 2014a).

This chapter introduces a unique Bahraini model of a western medical school branch campus. It explores the outcomes of this self-funded educational model—i.e., funded by the foreign partner, except for the land used, which was provided for free by the Bahraini government—in terms of its financial sustainability and its impact on education in the Gulf. While many western universities have been enticed by generous governmental incentives to set up branch campuses in the Gulf, the Royal College of Surgeons Medical University Bahrain (RCSI Bahrain) set up its branch campus in the Kingdom of Bahrain with an investment from the parent university in Ireland of nearly 100 million US dollars.

We will argue that the RCSI Bahrain model of an international branch campus can be described as a success, not only in financial terms for the parent University, but by the impact on the healthcare sector, and the higher education community, which includes its graduating students. This case analyzes this impact and questions why RCSI Bahrain has been successful along the dimension of a relatively large participation of domestic students (40%) compared to other models. The research for this case study is based on previous studies undertaken by Tawash, Cowman, and Anunciacion (2012); Hayes et al. (2013); Whitford and Hubail (2014) at RCSI Bahrain, national statistics and analysis of documentation. We also conducted a number of interviews to get a more complete picture of the functioning and outcomes of RCSI Bahrain.

Higher Education in the Gulf, a Shifting Landscape

Driven by knowledge economy ambitions, the demand for Western education in the Gulf has been met by the supply drivers of the changing models of universities in the West and the need to generate alternative funding opportunities. This globalization of education is characterized by neoliberal global market economic principles of competition and choice, which are often at odds with the more traditional philosophies of education for 'public good' (Steger and Ravi 2010). At the core of the neoliberal higher education model is the premise of competition, self-interest and entrepreneurship as tools for market growth. Green (1997) critiques education as moving towards being more for 'skill formation' rather than for 'citizen formation'. He discusses the fact that the traditional drivers for education such as equality and social justice have given way to capitalist principles of education to improve national competitiveness and directly links education with work. While Green's discussion is within the context of changes to education in older nation states, rather than the context of the newer nation states of the Gulf, the rhetoric of Gulf decision makers would indicate that this is also true of the Gulf economies (Donn and AlManthri 2010, Jones and Punshi 2013). This observation is echoed in literature that highlights the contradiction of the importance of skills formation in economies that are aspiring to compete on knowledge resources (Wheelehan 2012). Several national strategic documents of the GCC governments would indicate that skills formation and work based competencies are a priority for education in response to identified skills gaps in the labor market. Terminology like 'employability skills', '21st century skills' and 'work based skills' are used extensively in government documentation (Jones and Punshi 2013, Islamic Development Bank and International Finance Corporate 2011). These polices and strategies impact the labor market through support for internship programs, job placements and industry based qualifications as part of formal Higher Education, on job training, enterprise start up schemes (Jones and Punshi 2013). These schemes are often administered by purpose built government organizations that work very closely with policy makers and industry. In Bahrain, Tamkeen (labor market) fund is responsible for bridging labor market needs with education and up skilling the workforce.

With the expansion of private education institutions setting up in the GCC in the past decade, evidence suggests that the neo liberal market principles have very much dominated GCC government's education strategies (Donn and AlManthri 2010). However, there is a renewed recognition in Bahrain of the importance of building a strong indigenous education system (defined by a high participation of local citizens as students and staff) developed specifically to meet the unique local needs and utilizing internationalization of higher education in a more formal local capacity building role (Karolak 2012, Higher Education Council 2014a).

Researchers define internationalization as a series of dynamic processes of the transfer of people, programs, processes, curriculum, knowledge, ideas and quality standards across regional or national borders (Knight 2007). It incorporates program, policy and procedure (Knight 1997). However, internationalization can be seen as a double-edged sword when it comes to capacity building. On one hand there is significant opportunity for knowledge transfer and support in the development of national capacity, especially in areas of Research and Development (R&D) as stated by Vincent-Lancrin (2007) '*foreign educational services may actually be a means of accelerating the development of a national university system in the medium term*' (p.86). However, there is also the potential danger that importing programs and curriculum does not allow for the development of national human capability within higher education at local institutions (Knight 2013).

The United Nationals Development Program (UNDP) outlines the importance of capacity building as fundamental to their work within developing economies, especially when looking at developing knowledge resources. UNDP define capacity building as:

> '*Capacity is the ability of individuals, organizations and societies to perform functions, solve problems, and set and achieve goals. Capacity development entails the sustainable creation, utilization and retention of that capacity, in order to reduce poverty, enhance self-reliance, and improve people's lives. … Capacity development builds on and harnesses rather than replaces indigenous capacity' (United Nations Development Program).*

The true impact of local capacity building through importing international universities into the GCC can only be truly measured historically. As the locating of international branch campuses in the Arabian Gulf is a relatively new phenomenon, with a small number of branch campuses established in the early 1990's, approximately 5 across the Gulf, and the majority of branch campuses in the Gulf being founded in the 2000's (Willoughby 2008), one could argue that it is too soon to truly measure the high level capacity building impacts to any meaningful degree.

Authors Donn and AlManthri (2010) were among the first to investigate the impacts of globalization and internationalization policies on Gulf States. While they do not investigate capacity building specifically, they look at whether global education discourse has influenced the degree to which Western education (and labor market) policies have been 'borrowed' by Gulf governments. They discuss this 'policy borrowing' and its potential to lessen the purchasing countries motivation and capability to develop '*indigenous knowledge based educational developments*' (p.14). According to the aforementioned authors, this can lead the purchaser country to buy into subservience as imported expertise and knowledge often do not stay within the country and therefore threaten the opportunity to build local capability. Although strategic documentation from the various Gulf States would indicate a strong desire to develop the public education systems, it is also clearly stated that private higher education and forms of public and private partnerships will play a role in this development (Higher Education Council 2014a). There is also significant opportunity for stronger regional collaboration in utilizing internationalization strategies and building stronger indigenous education systems. So far, some of the Gulf States have set up various independent incentives and systems for attracting international branch campuses. However, there has been little coordinated funding and cross-border sharing of experience on educational projects, which might prove to allow greater regional benefit.

The knowledge-economy ambitions of the Gulf States are clearly outlined in various strategic documents (*for example* Bahrain 2030 Vision). However, it could be argued that this could pose an 'emperors new clothes' situation, as defining a country as having a knowledge economy is only possible when the outputs of that knowledge makes an impact and is tangible and measurable (Young 2012). Tangible outputs of knowledge can only occur when national absorptive capacity leads to publications, patents, R&D facilities etc. These outputs are the metrics that various innovation and competitiveness rankings use to measure knowledge. When looking at the development of societal knowledge, Wiig (2007) discusses that people are society's knowledge agents through both formal and informal education systems. He refers to the various types of knowledge within a society but states that those are explicit outputs and tacit transfer of knowledge that are fundamental to quality of life and societal good (Wiig 2007). This observation is supported by (Lundvall et al. 2009) who state that in the development of national innovation systems it is the interactions and interdependence of both the informal and formal elements within the innovation system that defines the outputs. This poses a challenge when measuring the impact and growth of knowledge as the formal metrics of impact, such as research publications and patents, do not include the informal elements within the system and that this therefore, forms an incomplete evaluation.

An understanding of the interaction between knowledge, education and society and between the policies and practices that underpin these interactions can only be considered within the complex cultural and societal nuances of the Gulf States. One of the cultural dimensions of the Gulf States can be described by what Trompenaars and Hamden-Turner (2004) calls 'particularism'. Members of particularism cultures favor human relationships over formal rules and regulations and look at friendships to help determine what is applicable and right in a given situation. This is as opposed to 'universalism' cultures where rules, values and standards take precedent over relationships (Trompenaars and Hamden-Turner 2004). This is an important consideration when discussing this case study as the 'fit' of RCSI to Bahrain was built on an understanding of particularism as being a dominant feature of Bahraini culture.

It is within this cultural context and an understanding of the limitations of measuring impact that this chapter will provide an overview of Bahrain and its healthcare sector, the RCSI Bahrain model, the impact that this branch campus has had on the Kingdom of Bahrain and lessons learnt. The wholly owned model of RCSI Bahrain is a unique example of an international branch campus in the Gulf whose significant financial investment in setting up a branch campus provides a case study in the establishment and operations of a branch campus.

The Kingdom of Bahrain and the Challenges in its Healthcare System

Economics and Demographics

The first decade of the 21st century was characterized by substantial economic and population growth for the Kingdom of Bahrain. Reaping the benefits of substantial investment in infrastructure (including a bridge connecting Bahrain to Saudi Arabia), education, healthcare and property construction, the Kingdom enjoyed rapid economic growth with an annual growth rate of approximately 5% between 2000 and 2012. This allowed substantial progress towards the diversification plans of the economy. Unlike some of its wealthier neighbors, Bahrain has only a modest energy endowment forming a negligible amount (0.03%) of the world total oil reserves compared to its neighbors Oman 0.3%, Saudi Arabia 16.1%, UAE 5.9%, Qatar 1.5% and Kuwait 6.1%. The gas reserves in Bahrain form 0.2% of the world gas reserves compared with Oman 0.5%, Saudi Arabia 3.9%, UAE 2.9%, Qatar 12% and Kuwait 0.9% (Al-Katiri 2013). The need to diversify the economy has become a key priority of the public agenda (Bahrain Economic Development Board 2010). In response, the Bahrain Government shifted investment into liberal market reforms designed to foster foreign direct investment and entrepreneurship. This has seen the impact of the energy production sector on GDP decrease from 40% in 2000 to 20%

in 2013, while the contribution from the financial sector rose from 14% to 17% and the contribution of the manufacturing sector rose 12% to 15%. The contribution from sectors including retail, construction, transport and communications also rose during this period (Bahrain Economic Development Board 2013).

The population of Bahrain is approximately 1.2 million people with 51% of this population being made up of expatriate labor. Of the Bahraini population, half are under the age of 25. With a large percentage of the nation's population being youth, Bahrain faces the same labor market challenges of its neighboring GCC economies, although to a lesser extent. These challenges include high youth unemployment figures, high dependence on expatriate labor, a gap between the labor market requirements and skills development of graduates from the higher education sector, low levels of local employee engagement (AON Hewitt 2013). However, Bahrain's labor market has followed a slightly different pattern as Bahrain enjoys some of the highest labor market participation rates of nationals in the Gulf at 31% of the labor force. This in contrast to the labor market participation rates of Saudi Arabia 22%, UAE 9%, Qatar 8% and Kuwait 18% (Hertog 2012). While there are a number of factors influencing the higher numbers of Bahraini nationals in the labor force, part of this is due to the labor market reforms early this decade that established a number of agencies and organizations focused on developing Bahrainis. These agencies are responsible for increasing Bahraini contribution to the labor market in terms of both numbers and productivity (Tamkeen). The two key government agencies with this mandate are Tamkeen and the Labor Market Regulatory Authority (LMRA). Tamkeen was set up to support Small to Medium Enterprises (SME) development through financing and skills development programs for Bahrainis. The LMRA was established to regulate and manage the registration of the workforce in Bahrain (World Economic Forum 2013). In addition to these agencies, the Bahraini government has announced in late 2014 the creation of a Labor Market Observatory that aims to balance labor market needs with education and training provision.

At the core of the Bahraini labor market reforms of the 2000's is the desire to ensure Bahrain remains competitive on a global scale. In the World Economic Forum 2014-2015 Global Competitive rankings Bahrain stands at 39th out of 144 economies. Bahrain ranks fifth out of the six Gulf countries, ahead of Oman at 62. While the Gulf countries are characterized by very high income levels, the Kingdom of Bahrain does not have the same wealth as some of its neighbors. Bahrain's GDP per capita is $27,435, compared with Qatar ($100,260), UAE ($43, 875) and Kuwait ($47,639) (World Economic Forum 2015). Yet, this relatively high income means Bahraini citizens enjoy a good standard of living that includes free access to education and healthcare.

Culturally Bahrain is considered more open and liberal than some of its neighbors. Western expats enjoy many of the same liberties they are used to from their home countries including liberal alcohol laws, dressing in Western clothing, women and men mixing and freedom of religion. This makes it an appealing place for expatriates to live (HSBC Expat Explorer 2014). Bahraini nationals tend to integrate with Western expatriates to a greater degree than some of the other Gulf States and it is not unusual for Westerners to have Bahraini friends. However, trust tends to be reserved for family groups or those who have built trust over time through relationship and it is common for those new to Bahrain to underestimate the importance of investing in relationships prior to being productive in business.

Healthcare in Bahrain

Bahrain's healthcare system is characterized by strong primary healthcare centers, three public hospitals and numerous private hospitals (Ministry of Health 2012). However, the national healthcare system is struggling to keep up with the rapidly increasing population. According to the World Health Organization, the GCC has low ratios of hospital beds per population at 21 per 10,000 residents, in contrast with 137 in Japan, 82 in Germany and 30 in the USA (World Health Organisation 2013). This current shortage, coupled with the continued population growth, means Bahrain is likely to be unable to meet expected future healthcare with its current capacity. Other healthcare issues characterizing the GCC and Bahrain include a high dependence on expatriate doctors and nurses, prevalence of certain diseases including type 2diabetes and variance in standards and procedures across the medical industry. The Bahrain government is well aware of these healthcare challenges and has developed a number of strategic initiatives to address them (Ministry of Health 2012).

Healthcare Education

There are three healthcare universities in the Kingdom of Bahrain: Arabian Gulf University; the College of Health Sciences; and RCSI Bahrain. Arabian Gulf University (AGU) was established as a coordinated GCC university. The aim was to start satisfying the needs of GCC countries in terms of specialists and experts in various occupational fields. AGU admitted its first medical students in 1982 from all six GCC countries. It has since expanded to offer postgraduate medical and science programs. Nurse training was first established in Bahrain in 1959. Between 1965 and 1975, other in-service training programs were initiated by the Ministry of Health e.g. laboratory and radiography technicians, pharmacy technicians, nurse-midwives, etc. The College of Health Sciences was established in 1976 under the governance of the Ministry of Health to provide training for nurses and allied health professionals. The College of Health Sciences has recently been moved under the University of Bahrain.

Within the GCC, there were a total of 27 medical universities in 2010 (Hamdy et al. 2010). Of these, 20 are less than 10 years old, many having graduated only small numbers of students. While some of these medical universities have international partnerships or international accreditations, only Weill Cornell Medical College in Qatar and RCSI Bahrain are campuses of foreign universities in the Gulf. Weill Cornell Qatar opened in 2002 as a partnership between Cornell University and the Qatar Foundation. The Cornell Board of Trustees provides oversight of the academic programs, whilst the Qatar Foundation provides and is responsible for the facilities. RCSI Bahrain opened in 2004 and is the only example of a medical university in the Arabian Gulf that has funded its facilities and programs entirely itself. RCSI Bahrain is fully owned and operated by the Royal College of Surgeons in Ireland, with an investment of close to one hundred million USD. For RCSI Bahrain this has led to a major commitment to see the project succeed.

RCSI and RCSI Bahrain

The Royal College of Surgeon's Ireland was founded in 1784 as a guild to practice surgery and to make provision for surgical education. Influenced by the surgical practices of France and the demand for medical professionals during the Napoleonic wars, RCSI rapidly developed its college. In 1977 The National University of Ireland established an agreement with RCSI Dublin that its graduates received a degree from its University. In 2010 RCSI Dublin was given the power to award its own degrees within the Irish education framework. RCSI Dublin offers an extensive range of undergraduate and post-graduate medical programs. It has long established relationships with hospitals in Ireland and around the world which offer student clinical placements, opportunities for graduates and research relationships. RCSI currently has 3,800 students and 952 staff across the world (Royal College of Surgeons Ireland Medical University Bahrain 2014).

RCSI Dublin's involvement in the Arabian Gulf began in the early 1990's with the Tabuk Saudi Arabia project, which involved supporting the management and operations of the Armed Forces Hospital at Tabuk. This project involved using on-line surgical training, which in 1995 was a relatively new phenomenon. This online surgical training was a joint venture between RCSI, Intuition Online Education and Harvard Medical School. During this time relationships with Bahrain were further developed.

As early as the 1970's Bahrainis were travelling to RCSI Dublin for postgraduate surgical training. Some of these graduates then went on to hold senior positions within the medical field in Bahrain, including Minister of Health. The relationship with Bahraini graduates led to RCSI Dublin staff being invited to Bahrain to offer post-graduate surgical training in Bahrain. By the 1980's RCSI were offering two postgraduate exams in Bahrain. By the late 1990's there was a strong relationship between Bahrain and RCSI Dublin with postgraduate training in surgery, family medicine and healthcare management being firmly established in Bahrain. In 2001 it was proposed that RCSI Dublin set up a campus in Bahrain. These discussions were formalized in 2003 with the Prime Minister of Bahrain visiting RCSI Dublin and inviting RCSI to establish a medical school in Bahrain. The Prime Minister offered a plot of land for the campus to be built, however all other costs for the establishment of RCSI Bahrain were funded by RCSI Dublin. A license was issued in 2004 to establish RCSI Bahrain and the first student intake was admitted in October 2004. RCSI Bahrain noble purpose states; "*Building on the heritage of the Royal College of Surgeons in Ireland which was founded in 1784, we will enhance health in Bahrain, the other GCC countries and beyond through endeavor, innovation and collaboration in education, research and service*" (Royal College of Surgeons Ireland Medical University Bahrain 2012).

The establishing of RCSI Bahrain was part of RCSI's broader internationalization goals. With a large number of students at RCSI Dublin coming from the Arabian Gulf and Malaysia, it was logical to establish branch campuses in these places. In 1995 RCSI Penang was established in Malaysia to deliver some postgraduate programs. In 2011 a campus was established in Perdana, Malaysia. This is a different model to RCSI Bahrain as it is owned by Malaysians with RCSI delivering the programs under contract. The Dubai campus was established in 2005 and delivers post graduate healthcare management programs.

RCSI's international reputation was validated by the 2015 Times Higher Education world university rankings, where RCSI was ranked within the top 50 universities in the world for international outlook and within the top 400 in the world. These rankings measure parameters for each university including teaching, research, citations, industry income and international outlook.

The Bahrain campus of RCSI is wholly owned and operated by RCSI Dublin. The leadership of the Bahrain campus embraced local Bahraini culture and invested in developing relationships to ensure RCSI made the desired impact. From its inception as a formal institution, RCSI sought the direction and guidance of Bahraini healthcare and higher education leaders to ensure the new institution added value to the Kingdom and complemented existing institutions.

RCSI Bahrain was set up with funding from the parent university at an investment close to $100 million USD. Students receive degrees from both RCSI Bahrain and the National University of Ireland. The same examinations and similar curricula are delivered in both Dublin and Bahrain. During its 10 years in Bahrain, 307 physicians and 547 nursing students have graduated from RCSI Bahrain (Royal College of Surgeons Ireland Medical University Bahrain 2014). These graduating students are from 36 countries with 60% being from Bahrain or other GCC countries. Clinical tuition and placements are within three governmental hospitals: Salmaniya Medical Complex (SMC); Bahrain Defense Forces Royal Medical Services Hospital (BDF) and the more recently opened King Hamad University Hospital (KHUH). RCSI Bahrain also has clinical tuition agreements with primary healthcare facilities in Bahrain. Distinctive of the RCSI Bahrain model is the large percentage of international students, faculty and relationships. As of 2013 approximately 40% of the student body is from Bahrain, 20% from other GCC countries and 40% from outside the GCC including Canada and USA (Royal College of Surgeons Ireland Medical University Bahrain 2014).

How the Relationship Works

RCSI Dublin works with RCSI Bahrain on all levels, including governance, finance, quality assurance and curriculum development. While this is seen as a key to the success of RCSI Bahrain, it has also provided a number of challenges. Some of these challenges could be described as normal challenges of branch campuses by internationalization expert Knight (1997) including autonomy, establishing structures and processes, contextualizing curricula, lack of infrastructure and support staff at branch campus and parent/branch communication. There are some areas of autonomy within some areas of the RCSI Bahrain campus, establishing structures and processes. However, there is less autonomy in other areas such as designing curriculum. In addition, the loss of infrastructure and support staff that are taken for granted in a parent institution can be challenging to Dublin staff who relocate to RCSI Bahrain.RCSI Bahrain is an independent constituted university, awarding its own degrees whilst awarding National University of Ireland degrees in addition. This somewhat confusing situation probably defines the governance and relationship between RCSI Bahrain and RCSI Dublin in other contexts. RCSI Bahrain is an operational business in terms of annual revenue. However, RCSI Dublin has invested large sums of capital. This independence, yet total dependence, underlies every area of the internal functioning of RCSI Bahrain. Externally, RCSI Bahrain is an independent university needing to establish its own recognition, accreditation and quality assurance. The linkage with the parent institution is helpful in this regard through its reputation. Nevertheless and quite appropriately, RCSI Bahrain has had to undergo all the quality assurance and accreditation procedures in its own right. Even more so, these are duplicated and triplicated related to the geographical position with RCSI Bahrain having to satisfy Bahraini, GCC and Irish requirements. In December 2014 RCSI Bahrain received accreditation from the Irish Medical Council. It followed an inspection visit by a delegation from the Irish Medical Council to RCSI Bahrain and its affiliated teaching hospitals. The accreditation applies to the University's undergraduate medical program and is for a period of five years. Accreditation is made on the grounds that the program adheres to the requirements, criteria and standards approved by the Irish Medical Council - these being the World Federation of Medical Education (WFME) standards, which are the highest standards available in medical education. This was a significant endorsement of the quality of RCSI Bahrain being regarded as of an international standard. In 2013 RCSI Bahrain underwent a quality review by the Bahrain Qualifications Quality Assurance (QQA) and received the highest confidence rating.

Part of the success of RCSI Bahrain in enabling X% of local students out of the graduating classes is due to the high levels of academic and language support offered to all students. As all instruction and examinations as are undertaken in English, RCSI Bahrain offers students formalized English language support. International students are also offered Arabic language lessons. Students at RCSI Bahrain take the same curriculum and examinations as students at RCSI Dublin and this is delivered using Bahrain based, predominately western, academic medical staff and local Bahraini medical professionals. The curriculum is developed through consultation between the international campuses and is benchmarked against the World Federation of Medical Education standards. Whilst the same curriculum is delivered in Bahrain as in Dublin, there has been the need to contextualize it to the local culture and healthcare system.

Since the same RCSI undergraduate medical programme is delivered in Bahrain, a crucial element of the coordination between the two campuses is to ensure that academic standards are exactly comparable. Students have the same lecture content and the same examinations. A major challenge has been to ensure that staff on both campuses are fully involved in curriculum development and examination preparation. Recruiting high quality staff in international settings is also a challenge.

Having high levels of interaction and exchange between Dublin and Bahrain is a crucial aspect of maintaining quality and equivalence across the sites. RCSI-Bahrain academics are represented in all major working groups in curriculum, assessment, research, evaluation and medical education throughout RCSI. Representation is also available at major academics committees and Boards in Dublin (Faculty Executive, Academic Council and the Medicine & Health Sciences Board). Contact is maintained by regular video-conferencing and on-site visits by Dublin staff to Bahrain. Similarly, Bahrain staff are facilitated to travel to Dublin for regular business, including attending committees, research events and faculty development-related activities. A further challenge is presented in ensuring the same learning outcomes are achieved in two countries with different cultures and healthcare systems (Whitford and Hubail 2014).

During the ten year development of RCSI Bahrain, many of these challenges have been overcome with RCSI Bahrain having now established its own structure and processes.

The Impact of RCSI Bahrain

In discussing the various stakeholder groups to international education, Knight (1997) suggests an analysis of the various 'sectors' affected by international education. Knight states the purpose of highlighting different sectors and the various stakeholder groups within the sectors is to illustrate that higher education is not the only group with a strong vested interest in international education. While this may seem obvious in this case study that is focusing on education in the healthcare industry specifically, it is still important to take a broader stakeholder perspective.

It is within the context of the challenges of measuring impact of a relatively new institution that the following section will discuss how RCSI has affected various sectors and stakeholder groups within Bahrain. Firstly we will discuss how RCSI has impacted the healthcare sector, then how RCSI has impacted the broader community within Bahrain, then the impact on the higher education and research landscape and finally, the impact on students themselves. While in reality these various stakeholders overlap in the categories outlined here, they will be discussed here separately.

Impact on the Healthcare Sector

The Government of Bahrain has a clearly outlined National Health Strategy that identifies a number of strategic priorities for the future of the sector. With an annual population growth rate of 7.4% between 2001-2010, it is recognized that Bahrain will need to increase the capacity of the healthcare industry to support the future needs of the growing Bahrain society. The Health Agenda Strategy of Bahrain (2012) states; '*the supply of health professionals from local and international academic institutions is not growing at the rate commensurate with the increased demand and expectations from the healthcare services*'. Based on population statistics and data from the Bahrain Ministry of Heath, it is forecast that by 2025 healthcare demand will increase by 130%. This demand in healthcare will need to be met by a substantial increase in supply in the healthcare workforce. It is estimated that Bahrain will need to find an additional 23,000 health professionals by 2030 to meet benchmarks (Ministry of Health 2014). To assist with meeting these labor market challenges, RCSI Bahrain has recently expanded the numbers of students that are enrolled in medicine (see Figure 2). This has been achieved without compromising quality as is shown by the improving exam results and lower attrition rates.

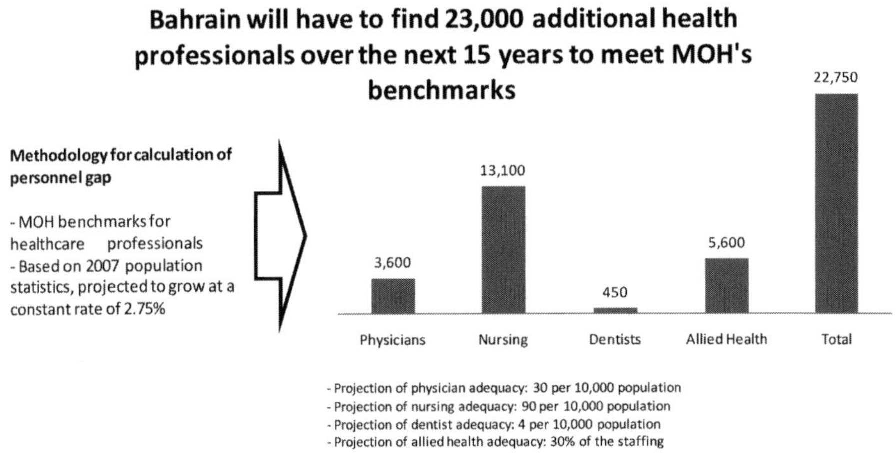

Figure 1: Developing Bahrain's Healthcare Workforce for Tomorrow, Planning Directorate, 2007. Source: Ministry of Health.

The above statistics highlight the importance of building local capacity within the sector to support future demand while containing costs. The establishment of RCSI Bahrain has gone a long way to contributing to an increase in the number of professionals to the sector. It is inevitable that part of the supply to the increasing demand for healthcare professionals in Bahrain and the Gulf will continue to be met by expatriates. According to the Ministry of Health, in 2012 Bahraini physicians made up 83% of the total number of physicians. However, expatriate nursing professionals make up 48% of the sector (Ministry of Health 2012) (See Figure 2). The rate of Bahraini nationals working as nurses in Bahrain is high compared with only 10% of nationals in Qatari and 8% of nationals in the UAE. However, the value of locally trained nurses with understanding of local languages and cultural customs is well recognized by the Ministry of Health who has clearly stated intent to increase the number of Bahraini nurses. With the current enrollment of Bahraini's into the RCSI Bahrain medical degree, it is expected that approximately 1,000 Bahraini doctors will enter the workforce in the next five years from RCSI Bahrain (Royal College of Surgeons Ireland Medical University Bahrain 2015).

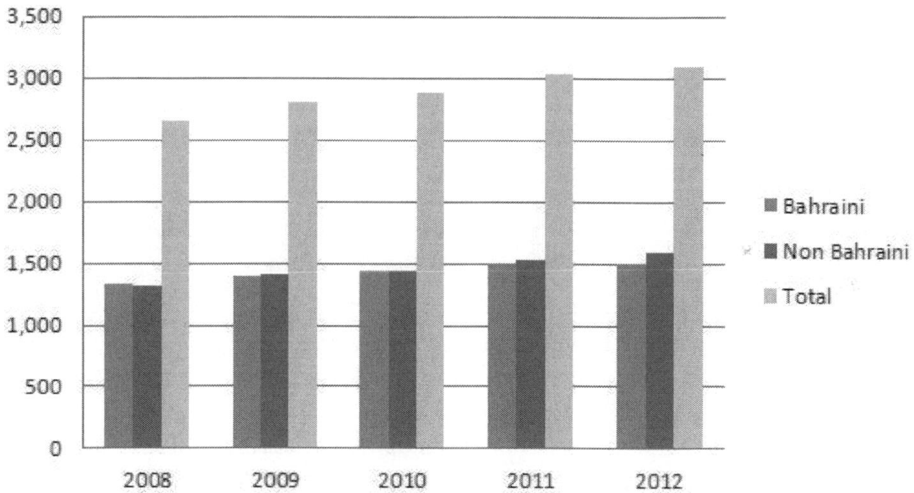

Figure 2: Number of Nurses in Bahrain. Source: Ministry of Health, Human Resources Manpower Statistics 2012.

RCSI Bahrain had its first intake into the nursing program in 2006. Since this first intake the numbers of Bahrainis entering the program have increased by over 150% (Cowman 2014). In addition, the number of male students entering nursing has increased by 25% (Cowman Bahrain, meaning that any Bahraini interested in developing their career needed to look for post graduate study off shore. RCSI Bahrain has been running an MSc in Nursing since 2009 and have agreed the development of the first Masters degree in Specialist Clinical Nursing in the region. The commencement date for the program is September 2015. With these post graduate qualifications, RCSI Bahrain is expecting to see an increase in clinical nursing research outputs.

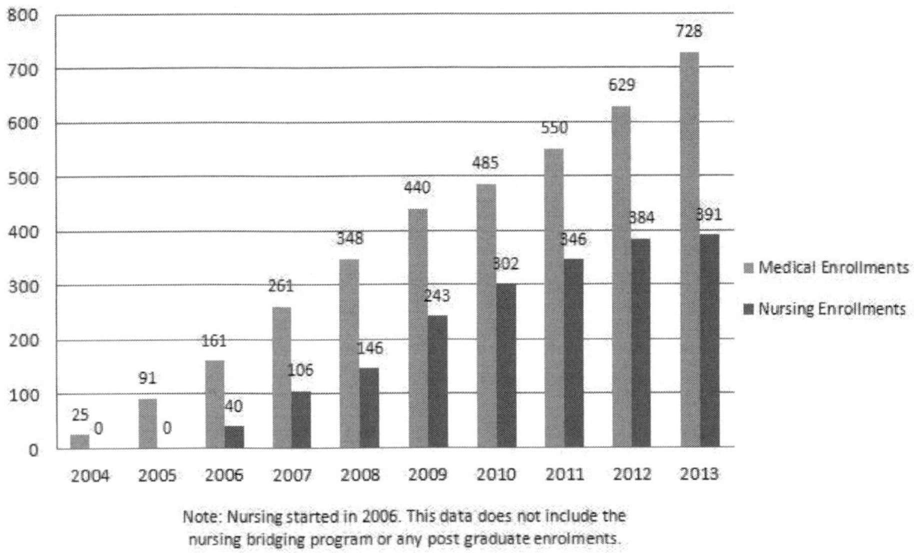

Figure 3: Enrollment to RCSI Bahrain for medical and nursing programs. Source: RCSI Bahrain Enrollment Office

Another key area of impact of RCSI on Bahrain is in the perception of nursing as a career choice. Traditionally nursing was not seen as a favorable career choice in the Gulf as it is perceived to be a low pay, low status job (Tawash, Cowman, and Anunciacion 2012). The increase in the number of both female and male Bahrainis entering RCSI Bahrain's nursing program could indicate the positive impact RCSI Bahrain has had on the perception of nursing as a career choice for Bahraini. RCSI Bahrain nursing lecturer, Ms Eman Tawash, undertook her PhD research in the promotion of nursing as a career choice. The study she conducted unveiled so far some of the reasons why it was not a favorable career choice and identified strategies to promote nursing to school leavers and parents. One of the initiatives she recommended was developing a promotional video to be used in Bahraini high schools to attract school leavers to nursing as a career choice. This recommendation was accepted and this video will be shown in all government and private high schools to encourage school leavers to enter the profession starting in 2015.

A key support mechanism for the increase in numbers of Bahraini nurses is the joint development program between Tamkeen and RCSI Bahrain. As mentioned previously, Tamkeen is a government agency responsible for supporting the growth of the private sector through the development of skills to improve productivity. This program provides fully paid scholarships to eligible Bahrainis to complete the nursing degree at RCSI Bahrain. Upon completion of their degree, graduates from the program are guaranteed employment at their sponsoring organization, usually a public or private hospital.

RCSI Bahrain's School of Nursing and Midwifery was recently awarded the honor of hosting the regional chapter of Sigma Theta Tau International, a nursing honor society. With 135,000 members in 90 countries, this membership allows Gulf regional nursing professionals access to international networks, and to evidence-based information on best practice and professional development. Heading the first Gulf chapter of Sigma Theta Tau allows RCSI Bahrain to be at the centre of the community of practice for nursing in the Gulf. This prestigious association gives RCSI a level of credibility and professionalism that will assist in RCSI's reputation as leaders in healthcare education.

In 2013, the RCSI Bahrain School of Medicine was invited to join the Global Education in Medicine Exchange (GEMx) affiliated to the Educational Commission for Foreign Medical Graduate (ECFMG) and the Global Health Learning Opportunities Collaborative (GHLO), affiliated to the Association of American Medical Colleges (AAMC). Both initiatives provide a network of collaborating institutions around the world, facilitating global mobility for the students pursuing international clinical or research electives (Royal College of Surgeons Ireland Medical University Bahrain 2014).

The nursing program's impact is extended to private industry with the outreach program to Saudi Arabian Oil Company (Saudi Aramco). Saudi Aramco, one of the largest companies in the world by turnover, is located approximately 45 minutes drive from Bahrain, just over the border in Dhahran, Saudi Arabia. RCSI Bahrain signed an MOU with Saudi Aramco to deliver their Bachelor of Nursing bridging program at Saudi Aramco's premises. The arrangement had RCSI Bahrain's staff delivering week-long lectures in Saudi Arabia for an initial cohort of 20 nurses. These 20 students are health team leaders and unit managers within Saudi Aramco completed the program in January 2015. In early 2014 Saudi Aramco set up a joint venture with John Hopkins University to establish Johns Hopkins Aramco Healthcare Company. The purpose of this company is to deliver healthcare and support the well-being of Saudi Aramco's 350,000 staff and their families. It is expected that John Hopkins Healthcare Company will become a centre of excellence for healthcare in the Arabian Gulf to complement RCSI Bahrain. John Hopkins University will contribute to this venture with consulting support and expert knowledge in clinical areas. John Hopkins Aramco Healthcare are also planning to offer nursing education and training programs. RCSI Bahrain is in discussions with John Hopkins Aramco Healthcare regarding collaborating on nursing education.

While it is too soon to assess the long-term impact on capacity building within the healthcare sector, Figure 3 shows that there is a growing number of good-quality medical professionals trained in Bahrain. With 40% of these students coming from Bahrain it is evident that RCSI Bahrain has made an impact in terms of career choice for high school students. While the focus of this chapter is on the impact of RCSI on Bahrain, it is worth mentioning the impact of the Bahrain campus on RCSI Dublin. The campus is in its 10th year of operation and the business plan is on target financially. Establishing RCSI Bahrain has cemented the place of RCSI as an international healthcare education leader (as evidenced by the Times Education Supplement rankings) and has provided a springboard for further international development, both regionally and internationally.

Impact on the Community

RCSI Bahrain's level of commitment to the Kingdom of Bahrain was beyond financial as outlined in their original mission statement '*the primary mission of the Medical School of the RCSI-Medical University of Bahrain is to inspire, educate and train competent and caring graduates who are well prepared to enter specialty training programmes and become leaders in their profession. We undertake these activities to internationally recognized standards of excellence for the people of Bahrain, the other GCC countries and the wider community* beyond' (Royal College of Surgeons Ireland Medical University Bahrain 2012). Community Engagement is one of three key strategic pillars for the institution. RCSI Bahrain's Community Engagement Office is responsible for identifying ways to give back and impact the broader Bahrain community (RCSI Annual Report, 2014).

Bahrain has some of the highest rates in the world of type 2 diabetes. RCSI Bahrain and the Bahrain Diabetes Society formed a partnership to develop a Pediatric Mobile Diabetes Unit, funded by Rotary Manama. Started in 2012this mobile unit visits government schools and teaches children about type 2 diabetes prevention and the importance of a healthy lifestyle. So far, over 1,200 children have attended these sessions. The diabetes mobile unit is run by RCSI Community Engagement team with the volunteer support of RCSI year three nursing students. This community program has been well received by the schools and by the Ministry of Health.

RCSI Bahrain Community Engagement Office has assisted with improving cardio pulmonary resuscitation (CPR) capability among residents. In 2011, staff and students ran CPR training sessions in Bahrain's largest shopping mall for over 300 people. In addition, RCSI Bahrain created an Arabic language video on how to conduct basic chest compressions. This video was has been viewed by over 11,800 people on the flights of Bahrain's national carrier, Gulf Air and over 7,000 hits on YouTube. There are plans to develop an Arab language mobile phone application to support CPR awareness.

Medical students can undertake an elective community outreach program as part of their training. This involves the students working on a number of different outreach programs. Staff and students lend their expertise to a number of other outreach programs including health awareness days and blood bank donations. RCSI Bahrain students also have opportunities to impact the community internationally. A small number of students and staff from RCSI Bahrain have travelled to Vietnam and Lesotho to partake in community projects there.

Impact on Higher Education and Research in Bahrain

The Higher Education Council (HEC) was formed in 2006 to regulate and promote the higher education agenda in Bahrain. All universities, including RCSI Bahrain, need to comply with the Higher Education Council regulations for operations and quality standards. The Higher Education Council has recently launched its National Higher Education Strategy and a National Research Strategy. RCSI Bahrain senior leadership have contributed to the development of both strategies and will have a significant role to play in their implementation. One key element of the National Higher Education Strategy is strategic access. The higher education sector must not just ensure equitable access to all who have ability, but it also must respond to human capital and future labor market challenges of the Kingdom of Bahrain. This is particularly relevant to the health sector in Bahrain, which has worrying supply and demand forecasts. Currently only RCSI Bahrain and Arabian Gulf University offer medical education and only RCSI Bahrain and the College of Health Sciences offer nursing. However, dentistry is not yet offered in Bahrain. The Higher Education Council is currently in dialogue with RCSI Bahrain to expand the current offering to include dentistry. This would represent significant progress to meet the expected demand for dentists in Bahrain.

Along with the rest of the Gulf States, the Kingdom of Bahrain is categorized as having moved from a factor driven to an innovation driven economy by the World Economic Forum (World Economic Forum 2015), due to its high income levels. While Bahrain scores well in areas of business sophistication, technology readiness, quality of institutions and education, Bahrain scores poorly on the innovation pillars of competitiveness (World Economic Forum 2015). In particular Bahrain scores low on the quality of scientific research institutions (87 out of 144). R&D spending is low at 0.04% of GDP and research outputs are negligible. With the lack of research being a systemic problem to the development of STI in Bahrain, the National Research Strategy has been well received. As outlined on the graphs below, Bahrain has witnessed a gradual increase in the number of research outputs, but as a share of overall GCC outputs, Bahrain's contribution has decreased.

Research Publications in Bahrain

The National Research Strategy has focused on capacity building through collaborative research relationships with local, regional and international partners (Higher Education Council 2014b). One of the strategy's main aims is to disseminate STI developments into commercial output and social benefit (Higher Education Council 2014b). Currently, Bahrain does not have the structures to commercialize research. This includes a lack of legal expertise (patent lawyers) and a lack of integration between higher education and industry. The National Research Strategy is addressing these issues with a plan to involve the broader innovation system into the implementation of the strategy.

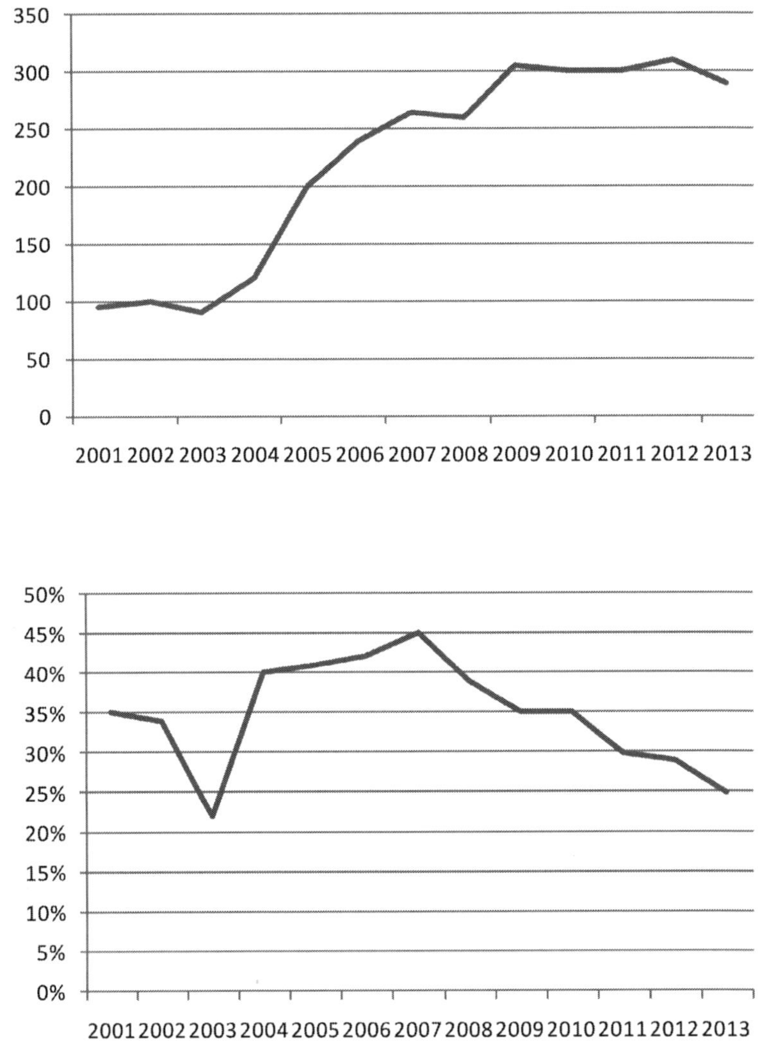

Figure 4: Evolution of Bahraini publications. Top panel: Total number of scientific publications; Bottom panel: Share of scientific publications in the region. Source: SCImago (2007). SJR SCImago Jouranl & Country Rank.

The National Research Strategy highlights a number of priority areas for research focus in Bahrain. One of these priority areas is research in Health Services and Public Health. RCSI Bahrain and Arabian Gulf University have recently signed a collaborative research agreement to create a $54,000 joint research fund to conduct medical research. This jointly funded initiative is the first of its kind in Bahrain and will go a long way to ensuring that both RCSI Bahrain and Arabian Gulf University and their respective international networks can mutually benefit Bahrain and the future of healthcare in Bahrain. Figure 5 outlines the structure in medical research that the National Research Strategy is intending to address:

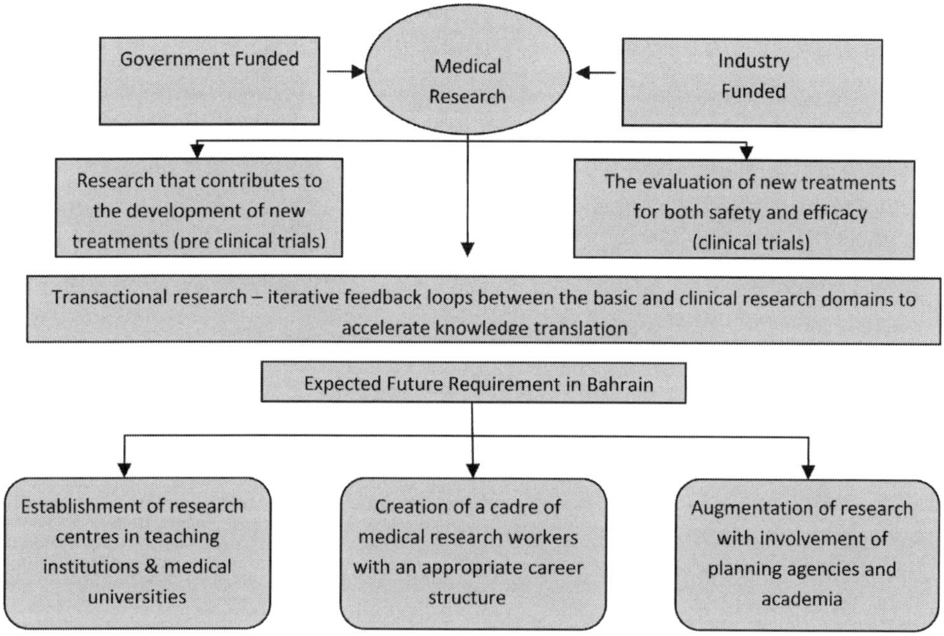

Figure 5: Structure in medical research that the National Research Strategy in Bahrain is trying to address. Source: Tamkeen Market Gap Study, Dunn & Bradstreet Research & Analysis, 2010.

Currently, across the higher education sector in Bahrain, less than 1% of the 32,527 students in higher education are studying at a PhD level (Higher Education Council 2014a). Currently only three universities, including RCSI, offer PhD programs in Bahrain. This extremely low percentage of students enrolled in research based degrees poses a concern for developing research capacity in Bahrain, in particular in the field of medicine and health sciences. With RCSI Bahrain's newly established Masters and PhD programs and their collaboration with Arabian Gulf University, it is expected that the future research outputs from RCSI Bahrain will make a significant impact on increasing Bahrain's research capability through involving Bahraini students and healthcare professionals in publishing collaborative research.

RCSI Bahrain has had 135 peer-reviewed papers published since opening in 2004. While this number remains modest, it is important to remember that as a new institution the focus has been placed on developing the appropriate structures and processes to facilitate teaching and learning. It is also important to note that many of these publications have been highly regarded by the medical community as evident by their impact factor, which measures the amount of times a publication is cited. 38 of these publications have an impact factor of 2 or higher, with 3 articles being 5 or above (Whitford 2014). This is an indicator of the high level of quality of the RCSI publications. Since 2001, the only publication from Bahrain in a high impact journal (defined as an impact factor greater than 15) has come from RCSI

Bahrain in the prestigious British Medical Journal (BMJ) (Benamer and Omran 2009). RCSI Bahrain has been also relatively successful in publications with undergraduates as 23 papers have been published so far by RCSI Bahrain faculty in collaboration with undergraduates. The involvement of undergraduates in research is voluntary but encouraged through providing funding (approximately $2500 per student) for students to spend their summer vacation engaged in research. Students must identify a faculty supervisor; write a competitive research proposal; and gain research ethics approval in order to participate. This mentoring of undergraduates into research serves as a role model for other higher education institutions in Bahrain, where the level of undergraduate involvement in research is currently very low.

The Higher Education Council intends to attract more international branch campuses and collaborative research arrangements to Bahrain (Higher Education Council 2014b). This is strengthened by the success of RCSI Bahrain as the Higher Education Council can point to a self-funded sustainable international branch campus operating model with ten years of history in Bahrain. It has directly contributed to the internationalization vision for higher education in the Kingdom of Bahrain, which is '*to position Bahrain as a regional hub for quality higher education producing graduates with the skills, knowledge and behaviors required to succeed in the global economy whilst contributing to the competitive growth of Bahrain*' (Higher Education Council 2014b). With 60% of RCSI Bahrain's students coming from abroad, RCSI Bahrain provides the country's best example of a successful international university in Bahrain. While other Bahraini universities receive international students, these are almost solely from other Gulf countries, usually Saudi Arabia. RCSI Bahrain has a very diverse student body with students from up to 36 different nationalities (Royal College of Surgeons Ireland Medical University Bahrain 2014). These international students enrich the learning environment at RCSI Bahrain and make contribution to the Bahrain economy.

As RCSI Bahrain is moving from an establishment to a development phase, they are well positioned to make a significant impact in research publication numbers. With 31 faculty having been involved in the 135 research publications, there is a level of experience in research and publication at RCSI that is among the highest within a higher education institution in Bahrain. There are three main research groupings at RCSI: medical education, oncology, and cardiovascular disease and risk factors. As part of the diabetes and oncology research RCSI Bahrain has a relationship with SBI Pharma, the only pharmaceutical company based in Bahrain.

Research is one of the pillars of RCSI Bahrain's 2014-2017 strategic plan. With research, in particular STI research, being essential to the development of a knowledge economy, RCSI will continue to contribute to Bahrain's economic development by being an important medical knowledge centre and example of research expertise in the Kingdom.

Impact on Students

At the core of RCSI Bahrain's impact is the student experience. As there have been only 854 graduates since 2010 it is too soon to measure long term impact after graduation.

However, there is evidence of RCSI Bahrain making an impact on Bahraini students as they are supported during their studies. It is well documented in literature that one of the key challenges for Gulf learners in higher education is the challenge of applying evidence based critical thinking skills that are common in western universities (Donn and AlManthri 2010). This has been a challenge for RCSI Bahrain as many of the Bahraini and GCC students entering their studies have come from a schooling background based on rote learning and not critical thinking. RCSI Bahrain has sought to address these issues in several ways. First, through a tiered admission system as some students have to carry out a Foundation year prior to the five year medical course; secondly, through a Foundation Year designed to deliver a curriculum concentrating on basic sciences and language ability, but also developing study skills, use of ICT, critical thinking and adapting to university life; thirdly through a pre-Foundation Year program (the Medical Commencement Program) as a university preparatory year; and fourth, through identifying students with poor English language ability throughout the whole course and offering remediation. The university has developed an active student support service with particular emphasis on development of study skills. Developing critical thinking and professional attitudes are more of a challenge as these attributes are probably best developed in earlier life. However, the curriculum has been designed to include summative assessments that address team working, reflective practice, ethical issues and some aspects of professionalism.

The important role of student support and the embedding of critical thinking skills has also been identified by the only other branch campus medical school in the Gulf, Weil Cornell in Qatar. This Ivy League medical school has graduated 181 of high caliber medical professionals from 7 graduating classes from 2008 to 2014, although the actual number of Qatari graduates among the first few cohorts of this program has been around 27%.

World Class Medical Education

As mentioned previously, medical students at RCSI Bahrain sit exactly the same examinations as those in Dublin. The exams are written jointly by faculty in both campuses. This enables comparison between the campuses. What has been discovered is that students entering RCSI Bahrain perform less well in examinations in their first two years than students in Dublin but by the final year of medical school there is no difference between students at the two campuses. This is particularly true for students from the Gulf region. We can conclude that the approach of RCSI Bahrain is 'adding value' to the students who choose to attend, particularly those who have remained within their own cultural setting. In addition, our top students also perform well. RCSI Bahrain students are eligible to sit the top performance examinations of the National University of Ireland. In 2014, two students from RCSI Bahrain won the National University of Ireland awards for excellence. These awards are designed to honor the top performing students in Humanities and Science from across all Irish Universities. This was the first time students from RCSI Bahrain won this prestigious award. Additional data exists from licensing examinations. RCSI Bahrain candidates for the Bahrain Medical Licensing Examination (BMLE) (sat at the end of the one year internship) have maintained an outstanding 100% pass rate as compared with the overall national pass rate for this examination of 40%. RCSI Bahrain graduates for the US Medical License Examination (USMLE) have a 75-80% pass rate, equivalent to that for foreign graduates of 76%. The average score for RCSI Bahrain graduates in the Canadian licensing examination is 350, compared to 271 for all foreign graduates sitting this examination. RCSI Bahrain and Kaplan Medical, a test preparation organization, recently signed an agreement, which will allow Kaplan Medical to offer its USMLE (US Medical Licensing Exam) step 1 at the RCSI Bahrain starting the summer of 2014. This is a lecture programme which includes live lectures offered by Kaplan's US medical faculty with the quality, structured, interactive experience normally only offered in Kaplan's American centres. This allows RCSI Bahrain students a chance to improve their pass rates on the USMLE licensing exam.

RCSI Bahrain is still a young institution but figures thus far indicate that nearly 100% of Bahraini medical graduates are choosing to stay in Bahrain to carry out their internship. This provides Bahrain hospitals with high quality national, locally trained junior doctors. In addition, the majority of RCSI Bahrain graduates from other GCC countries also stay in Bahrain for internship. They are joined by a group of RCSI Bahrain graduates from other countries. While it is too soon to quantify the impact that this has on improving local capacity, anecdotal evidence from the employers suggests that RCSI Bahrain graduates are very well received into the local hospitals, particularly for their professionalism, communication skills and general clinical approach.

The student experience at RCSI Bahrain is a reflection of the faculty. The majority of the faculty comes from Western countries. With Bahrain being considered a desirable expatriate location, the turnover of expatriate staff is relatively low (approximately 5% per year). This provides further sustainability to the students' experience at RCSI Bahrain. It also provides further sustainability to the long-term impact of RCSI in Bahrain as these staff members build both professional and personal relationships in the Kingdom, both with students and other professionals. It is from these relationships that a number of the community engagement initiatives have developed.

Challenges and Lessons Learnt

RCSI Bahrain has proved to be a successful model of an international western university branch campus in a GCC country in terms of the quality of training of nurses and doctors, the degree of participation of Bahraini nationals, and the overall contribution to the provision of healthcare. This success can be measured financially and through its early impact in Bahrain on the healthcare sector, involvement in the community, higher education, research and student outcomes. RCSI Bahrain is financially secure and is well on target for paying off the loans incurred in establishing the university. As a result, RCSI Bahrain is considering expansion of their programs, and is currently examining the feasibility of offering dentistry as part of their programs. In this section we look at the factors that may have contributed to this success.

Relationships and 'Fit'

Our central thesis is that RCSI Bahrain is a good 'fit' with the Kingdom of Bahrain in terms of culture, finance, and educational standing. We would contend that the suitability of the 'fit' between any western higher education institution and the local environment will be a sound predictor of success.

As discussed previously, Bahraini culture is particularistic because it places a high priority on relationships, including in the business world (Trompenaars and Hamden-Turner 2004). The existence of a significant number of RCSI Dublin alumni in Bahrain, the long history of postgraduate teaching from RCSI into Bahrain and an understanding of the relational nature of Bahraini culture by an Irish institution that shares aspects of this culture have all contributed towards both the initial establishment and ongoing success of RCSI Bahrain. This strong relationship has also allowed access to healthcare facilities in Bahrain for clinical training and placements and led to a commitment from RCSI Bahrain to be engaged with and serve the local community and healthcare needs. This engagement has been a critical part of the success of RCSI in Bahrain.

Relatively speaking, Bahrain does not share the same level of affluence (nor natural resources) as some of its neighbors. The willingness of RCSI to invest into higher education in Bahrain was necessary in this context but also further built the relational partnership with the Kingdom of Bahrain. In addition, it has led to an increased institutional commitment to see RCSI Bahrain succeed; a commitment that may often be less apparent when the only cost of failure is loss of prestige.

The relationships between RCSI and Bahraini partners have also been essential to avoid potential challenges that may have threatened RCSI in Bahrain. The mutual respect that has been built over a long period of time between RCSI Bahrain and local partners has established a platform that has been and will be able to withstand potential conflicts. The strength of local relationship is essential to mitigate risk to any international university considering working within the Gulf. As loyalty is seen as a highly desirable characteristic among the culture of the Gulf States (Trompenaars and Hamden-Turner 2004), RCSI's financial and personal commitment to Bahrain has been viewed favorably by Bahraini partners in the face of any challenges.

Academic 'Fit'

The challenge of institutional 'fit' into the local context may be even more important on an academic level. RCSI has good standing and international rankings (ranked between 350-400 in the Times Higher Education rankings) but is not classed as a top tier institute. It does, however, rank above other Bahraini higher education institutions. The advantage to Bahrain of this fit is that the university has experience and expertise in "adding value" to students and as a result is able to take students from the region and equip them to be fit for purpose, i.e. competent healthcare professionals. At the same time, RCSI is setting a standard for local higher education providers to aspire to. Consideration into the equivalency of student learning and student assessment has been a critical part of ensuring the same quality standards are maintained between RCSI Dublin and RCSI Bahrain. This included ensuring there are no compromises in student selection. However, due to the different levels of the schools systems between Ireland and Bahrain, some students are required to undertake a foundation study year prior to being admitted to the RCSI Bahrain degree program. The foundation year is a rigorous experience that supports the student to progress successfully into the medicine program.

Parent Campus Staffing

Some other key lessons learnt from the experience of RCSI in Bahrain include the importance of staff from the parent university being 'on the ground' for a period of time to ensure academic rigour and stability. Parent university academic standards and policies and procedures need to be contextualized to ensure standards are maintained and knowledge transfer to local staff occurs. This is also very important when considering engagement with Bahraini community and society. The long-term expatriate staff at RCSI Bahrain have integrated in various ways with Bahrain partners and contributed to the institution's success through this engagement and the relationships formed.

The main learning point from the RCSI experience in Bahrain is that responding to an invitation to establish an international higher education institution must be preceded by an understanding of the local culture, regulatory arrangements and needs. If there is a good 'fit' between the local culture and needs and that of the foreign institution, success is more likely to follow. Initial relationships can start at the STI grass roots level by engaging with collaborative research where there is a significant opportunity for immediate impact.

References

Al-Katiri, Laura. 2013. Energy Sustainability in the Gulf States: the why and the how? Oxford University: The Oxford Institute for Energy Studies.

AON Hewitt. 2013. Quadurat Report. Dubai: AON Hewitt.

Bahrain Economic Development Board. 2010. Bahrain 2030 Vision. Bahrain.

Bahrain Economic Development Board. 2013. Bahrain Economic Yearbook. Bahrain: Bahrain Economic Development Board.

Benamer, Hani, and Bakoush Omran. 2009. "Arab Nations Lagging behind other Middle Easter Countries in Biomeical Research: A Comparative Study." *BMC Medical Research Methodology*:9-26.

Cowman, Seamus. 2014. "Editorial." *Journal of Advanced Nursing* 70 (11):2417-2418.

Donn, Gary., and Yahra AlManthri. 2010. *Globalisation and Higher Education in the Arab Gulf States*. Oxford: Symposium Books.

Green, Andy. 1997. *Eduction, Globalisation and the National State*. Basingstoke: Palgrave.

Hamdy, H., A. W. Telmesani, N. Al Wardy, N. Abdel-Khalek, G. Carruthers, F. Hassan, S. Kassab, M. Abu-Hijleh, K. Al-Roomi, K. O'Malley, M. G. El Din Ahmed, G. A. Raj, G. M. Rao, and K. Sheikh. 2010. "Undergraduate medical education in the Gulf Cooperation Council: a multi-countries study (Part 1)." *Med Teach* 32 (3):219-24. doi: 10.3109/01421590903389108.

Hayes, A, C Holden, D Gaynor, B Kavanagh, and S Otoom. 2013. "Bridging the gap: A program to enhance medical students' learning experience in the foundation year." *Bahrain Medical Bulletin*:35.

Hertog, Steffen. 2012. "National employment, migration and education in the GCC. The Gulf Region: economic development and diversification." In *A comparative assessment of labor market nationalisation policies in the GCC*, edited by Steffen Hertog. Berlin: Gerlach Press.

Higher Education Council, Bahrain. 2014a. National Higher Education Strategy. Bahrain: Higher Education Council Bahrain.

Higher Education Council, Bahrain 2014b. National Research Strategy. Higher Education Council.

HSBC Expat Explorer. 2014. "Expat Explorer." Accessed 30 NovemberIslamic Development Bank and International Finance Corporate. 2011. Education for Employment; Realiing Arab Youth Potential.

Jones, David., and Rudhika Punshi. 2013. *The Paradox of Plenty*. Dubai: Motivate Publishing.

Karolak, Magdalena. 2012. "Bahrain's Tertiary Education Reform: A step towards sustainable economic development." *Revue Des Mondes Musulmans et de la Mediterranee* 131:163-181.

Knight, J. 1997. "Internationalisation of Higher Education a Concetual Framework." In *Internationalisation of Higher Education in Asia Pacific*, edited by J. Knight and H. de Witt.

Knight, J. 2007. "Cross Boarder Tertiary Education: an introduction." In *Cross Boarder Tertiary Education: A way towards capacity*, edited by OECD, 21-46. Geneva: Organisation for Economic Cooperation and DevelopmentKnight, Jane. 2013. "Crossboarder Education in the Gulf Countries; Changes and Challenges." In *Education in the Broarder Middle East: Borrowing a Baroque Arsenal*, edited by Gary and AlManthri Donn, Yahra. Oxford: Syposium Books.

Lane, Jason E. 2011. "Importing Private Higher Education: International Branch Campuses." *Journal of Comparative Policy Analysis: Research and Practise*:367-381.

Lundvall, Bengt-Ake, K Joseph, C Chaminade, and J Vang. 2009. *Handbook of Innovation Systems and Developing Countries*. Edited by Bengt-Ake. Lundvall, K. Joseph, c. Chaminade and J. Vang. London: Mixed Source.

Ministry of Health, Bahrain. 2012. "Human Resource Statistics 2012." Government of Bahrain.

Ministry of Health, Bahrain. 2014. Bahrain Health Agenda. Bahrain.

Royal College of Surgeons Ireland Medical University Bahrain. 2012. RCSI Bahrain Strategic Plan 2012-2017. Bahrain.

Royal College of Surgeons Ireland Medical University Bahrain. 2014. Annual Report.

Royal College of Surgeons Ireland Medical University Bahrain. 2015. Enrollment Statistics. edited by Royal College of Surgeons Ireland Medical University Bahrain. Bahrain.

Steger, M., and R. Ravi. 2010. *Neoliberalism: a very short introduction*. Oxford: Oxford University Press.

Tawash, E, S Cowman, and E Anunciacion. 2012. "A Triangulation Study: Bahraini Nursing Student's Perceptions of Nursing as a Career." *Journal of Nursing Education and Practise* 3.

The Times. 2015. "The Times World University Rankings." Accessed 08 November.

Trompenaars, A, and C Hamden-Turner. 2004. *Managing People across Cultures*. Oxford: Capstone.

United Nations Development Program. Accessed 5 March.

Vincent-Lancrin, S. 2007. "Developing Capacity through Cross Boarder Tertiary Education." In *Cross Boarder Tertiary Education: A way towards capacity*, edited by OECD, 47-102. Geneva: Organisation for Economic Cooperation and Development.

Wheelehan, L. 2012. "The Problem with Comptency-based Training." In *Educating for the Knowlege Economy; Critical Perspectives*, 152-166. London: Routedge.

Whitford, D. 2014. *10 Years of Reseach at RCSI Bahrain*. Bahrain.

Whitford, D, and AR Hubail. 2014. "Cultural sensitivity or professional acculturation in early clinical experience?" *Medical Teacher* 36 (11):951-957.

Wiig, K. 2007. "Effective societal knowledge management." *Journal of Knowledge Management* 11 (5):141-156.

Wildavsky, B. 2010. *The Great Brain Race: How Global Universities are Reshaping the World*. Princeton: Princeton University Press.

Willoughby, J. 2008. "Title." Working Paper Series.

World Economic Forum. 2013. Arab World Competitiveness Report. World Economic Forum

World Economic Forum. 2015. The Global Competitiveness Report 2015-2016

World Health Organisation. 2013. World Health Statistics.

Young, M. 2012. "Education, globalisation and the 'voice of knowledge'." In *Educating for the Knowlege Economy? Critical Perspectives*, edited by H Lauder, M Young, H Daniels, M Balarin and J Lowe, 139-151. London: Routledge.

About the Contributors

Laura Diaz Anadon is Visiting Senior Lecturer in the Department of Science, Technology, Engineering & Public Policy at the University College London and Assistant Professor at the Kennedy School of Government at Harvard University. Professor Anadon's research focuses on innovation institutions, energy and technology policy, managing uncertainty, and the policy implications of the interdependencies of resources. She has advised policy makers internationally and worked as a consultant for various international organizations. She was co-editor of the 2014 Cambridge University Press book *Transforming U.S. Energy Innovation.*

Jennifer DeBoer is an Assistant Professor of Engineering Education at Purdue University. Her research group is motivated to understand how technology and policy tools can promote equity and success for diverse engineering students around the world. She is the recipient of a 2015 National Science Foundation Early CAREER Award for her work evaluating and improving online courses for engineering undergraduates from diverse backgrounds.

Aruba Khalid is a Dubai based economic researcher and has worked at the UAE Ministry of Foreign Trade, as part of the UNDP Capacity Development Project, and at INSEAD Abu Dhabi Innovation Policy Initiative. Her key research interests include GCC labour market structures and science technology and innovation policy in the GCC.

Ashley Ater Kranov is President of the consulting firm Global Professional Skills Assessment. She and her team of consultants partner with faculty in computer and engineering programs in the Middle East to ensure that their curricula are aligned with industry needs and quality assurance organizations. She is also adjunct associate professor of Electrical Engineering at Washington State University in the USA.

Annamarie Lawrence is a lecturer in the Business School at Bahrain Polytechnic, a government owned higher education institution with a focus on industry integration and 21st century skills. Prior to Bahrain Polytechnic, Annamarie worked as a consultant on various multi-stakeholder economic development/education projects for the New Zealand government. Annamarie has also had management consulting experience in the private sector while working for PWC in Sweden and Bain & Co in London. Her research interests are in the role of education policy within developing innovation systems.

Khaled Machaca serves as Associate Dean for Research of Weill Cornell Medicine-Qatar (WCM-Q) since 2008. In that capacity he oversees the academic, financial, operational and compliance aspects of the research effort. This included establishing the administrative infrastructure; centralized core laboratories; and faculty recruitment & retention. His research interests are focused on Ca2+ signaling in physiological and pathological processes.

Cameron Mirza is the Director of Strategy and Projects at the Higher Education Council Kingdom of Bahrain. For the last four years he has been the lead person responsible for reform of the higher education sector and responsible for a nations improvement in seven out of eight indicators in one year for innovation ranking in the global competitiveness report produced by the world Economic Forum in 2015. Prior to that Cameron spent several years working in central government in the UK leading on high profile projects within education and social reform.

Sebastian Pfotenhauer is an Assistant Professor of Innovation Research at the Munich Center for Technology in Society (MCTS) and the TUM School of Management, both at TU Munich. He was previously a lecturer and research scientist at the MIT Technology & Policy Program and Engineering Systems Division, as well as a Fellow at the Harvard Program for Science, Technology and Society. His research interests include national, regional, and institutional innovation strategies; imaginaries of innovation; the global circulation of innovation models; and complex international science, technology and innovation partnerships for capacity-building. He has consulted for various governments and international organizations.

Javaid Sheikh is an internationally renowned medical executive and creative thought leader in global academic medicine. During his tenure as Dean of Weill Cornell Medicine-Qatar (WCM-Q) since 2010, Professor Sheikh has pioneered innovative educational and research programs enabling WCM-Q to become widely acknowledged as a leading institution preparing "Global Physicians" for the 21st century.

Afreen Siddiqi is a Research Scientist at the Massachusetts Institute of Technology and a Visiting Scholar in the Science, Technology, and Public Policy Program at Harvard Kennedy School. Her expertise is in complex socio-technical systems analysis, and her current work is on science and technology policy, and water, energy, and food security in developing regions. She has worked with leading companies in oil and gas, instrumentation, and aerospace sectors, and serves in an advisory role in international research institutions.

David Whitford is a Dublin-based Professor of General Practice at the Royal College of Surgeons in Ireland. He spent 8 years at RCSI Bahrain as Professor of General Practice and Head of the School of Postgraduate Studies & Research. During this time, he established the community-based teaching in the new medical school and encouraged the development of research. Prior to his posts in RCSI, he practised for 16 years as a General Practitioner in Newcastle upon Tyne, UK. His research interests are in Type 2 diabetes and the impact of socio-economic deprivation on health.